Quality Hand Soldering
& Circuit Board Repair

Third Edition

Quality Hand Soldering & Circuit Board Repair

Third Edition

H. (Ted) Smith
Quality Soldering Technology

DELMAR

THOMSON LEARNING™

Quality Hand Soldering & Circuit Board Repair
by H. (Ted) Smith

Business Unit Director:
Alar Elken

Executive Editor:
Sandy Clark

Senior Acquisitions Editor:
Gregory L. Clayton

Development Editor:
Jennifer A. Thompson

Executive Marketing Manager:
Maura Theriault

Channel Manager:
Mona Caron

Executive Production Manager:
Mary Ellen Black

Senior Project Editor:
Chris Chien

Production Editor:
Stacy Masucci

Art/Design Coordinator:
David Arsenault

Marketing Coordinator:
Paula Collins

For permission to use material from this text or product,
contact us by
Tel   (800) 730-2214
Fax   (800) 730-2215
www.thomsonrights.com

Library of Congress Cataloging-in-Publication Data
ISBN 0-7668-1410-6

NOTICE TO THE READER

Contents

Preface

When it comes to soldering, most technicians and electronic equipment assemblers—the people who need to know about this particular task—are the last ones to find out how it should be performed. People everywhere have failed to identify this as a critical area of study for technical people concerned with soldering, reworking, or repairing printed wiring boards.

The following article appeared in the *Ottawa Citizen* in August 1984 and shows just one documented example of why we need to become much more familiar with the requirements for good soldering skills.

BAD SOLDERING DOWNED 767, WITNESS TELLS INQUIRY

WINNIPEG (CP)—A defective soldering job probably caused the fuel gauge system on a Boeing 767 to fail, a problem that lead to the emergency landing of the aircraft near Gimli, Manitoba, investigators into the incident were told Thursday.

Diane Rocheleau, an analyst with the Aviation Safety Bureau, said tests showed there was insufficient contact between two soldered parts. Rocheleau testified that this caused the fuel measuring system to function intermittently.

A mistake in the manual calculations resulted in the plane leaving with only half the fuel needed to reach Edmonton but the jet made a safe emergency landing on an abandoned airstrip after its tanks ran dry somewhere over northwestern Ontario.

A film entitled "Falling from the Sky—Flight 174" was made based on this occurrence. A flight simulation program on the problems incurred during the flight was produced and pilots have had a great deal of difficulty successfully completing the program.

DEFINITION OF HIGH RELIABILITY SOLDERING

The soldering technique whereby the probability of obtaining perfect metallic joining, product cleanliness, and optimum electrical conductivity without damage to components or equipment has been statistically proven (ANSI/IPC).

Quality Hand Soldering and Circuit Board Repair, Third Edition, provides critical information for anyone who wants to learn about the process of manual soldering and related assembly aspects. If the suggested techniques and methods are adhered to, the result will be a product of high quality and reliability. The text provides everything necessary to carry this out whether in the manufacturing or service sectors of the electronics industry. The second portion of the handbook teaches technicians to make quality, reliable board repairs.

H. (Ted) Smith is an experienced course instructor, teaching throughout Canada since 1982. He has taught such courses as High Reliability Soldering, Mil-Std-2000, Repairing Circuit Boards, and Surface Mount Technology. In response to various companies' requests, he has developed and led one-day and longer courses geared to specific needs. He has also offered seminars on electrostatic discharge.

CHAPTER 1 General Soldering Information

Objectives

After studying this chapter, you should be able to:

- List the five steps of preparation for soldering.

- Describe how to properly tin a soldering iron.

- Describe how to care for your soldering iron.

- List the types of flux that should be used for electronics and why.

- Explain the term *eutectic*, what makes it unique, and the percentages of the two metals involved.

CLEANING

A good soldering technician observes the following stages of preparation for each job:

1. Cleaning all components, circuit board, tools, and materials to be used for the soldering process.

2. Selecting the flux.

3. Determining the heat to be used and length of time to do the job, which are based on the thermal mass of the parts to be connected.

4. Selecting the solder.

5. Choosing the flux remover.

Ask an experienced soldering technician, "What is the most important task to perform before soldering?" Many technicians, even those who have been soldering for forty years, will typically answer, "heat," "iron tip," "solder." They generally miss the most critical task of all: **cleaning.**

Clean the soldering iron tip, component lead or wire, the item that the component is being soldered to (board or terminal), tools being used to form the wires/leads, and even the solder itself.

Cleaning the Soldering Iron Tip

The soldering iron tip should be bright silver with no flux residue or solder on it. Any major buildup of oxide on the tip is removed by wiping the tip on a damp sponge before applying it to the area to be soldered. This shocking action steams off the oxide and leaves the tip pristine and in the proper condition for soldering. To start, you need the correct soldering iron for the job. For the majority of electronics work, this means a 25-to-30-watt pencil-type iron with variable heat control. This makes it easy to ensure that the proper temperature is used for the work at hand. Soldering guns or irons with magnetically controlled heaters can possibly damage the very tiny and sensitive integrated circuits or ICs because of the electromagnetic fields radiating from these types of irons.

Tip Maintenance

If a soldering iron tip does not have a thin consistent layer of solder over the entire surface, the tip has not been properly tinned. If the iron tip is not properly tinned, start with a cold iron, turn the heat on, and hold the flux cored wire solder against the tip as it heats up. Wipe off the excess solder, then shock the tip on the sponge before soldering. Do not wipe the excess solder, burned flux residue, and other contaminants onto the sponge. The purpose of the sponge is to shock the iron. If you keep dumping your excess solder, burned flux, and other residue onto your sponge, the sponge quickly becomes useless. Every time you then touch the sponge, you pick up the dirt you put there earlier. This adds contamination to the solder connection. The sponge should remove the thin layer of oxide that builds up when the iron is heated. Find another means and another place to remove the dirt from your iron. One method is to use a paper or cotton wipe, ones that will not shred and leave particles behind, and very gently wipe the dirt from your iron. Then shock the iron tip by touching the dampened sponge. The newest sponges have holes in them so that the excess solder, flux debris and other unwanted residues are not left on the surface of the sponge. The contaminants now have a place to go instead of being left on the sponge.

Practical Hint

When you are not using your iron, make sure you leave a large lump of solder on the tip. This maintains the tinning on the tip, and the tip will last much longer. Many technicians mistakenly clean the tip before they put the iron into the holder. Leave the solder on the tip to protect it.

Caution

Where solder has been left on the tip to protect it, it is not a good idea to flick or bang the iron to get rid of the excess solder. You could end up being burned or worse by the flying solder. Use the proper sponge or other means to remove it.

Board Cleaning

In a manufacturing facility, a relatively clean board is generally available, but this should not be taken for granted. If boards have been stored without protection against oxides and other airborne contaminants, cleaning may be required before you do any soldering. Wire terminals may need to be pretinned to remove oxides before a wire is installed. Dirt films on metals may consist not only of oxides, but also sulfides, carbonates, and other corrosive materials from the environment. These will hinder solder flow or wetting of the solder onto the surfaces being soldered.

Component Leads and Wire Cleaning

Component leads should be tested periodically for solderability. Take items from stock at random and test them to ensure that problems will not be encountered when components are installed onto the board. If necessary, re-tin the leads, then clean them off. Wire, tinned by hand or by solder pot, should have the burned flux residue removed. If this residue is not removed, this contaminating material will be included in an unreliable connection. Clean the wire with a liquid cleaner. Items such as a pink eraser, steel wool or similar types of cleaning tools are not a good idea. The eraser leaves a gun residue which you now have to remove and steel wool could actually remove the tinning, etc. Some technicians feel that the heat of the soldering iron cleans off the area to be soldered. This is a very common misconception. Some of the techniques used actually increase the oxidation rate. (This will be discussed in Chapter 2.) Make sure that everything you use or solder is **clean**.

FLUX

A second very important item in preparing to solder is the flux. Flux has a very definite purpose: It prevents oxidation and removes the thin layer of oxide and the atmospheric gas layer from the area to be soldered. When flux is applied to the area, it permits the solder to flow, or wet, smoothly and evenly over the surface of the lead, wire, or pad being soldered. It also improves the flow of heat, resulting in faster heating of the items or area being soldered.

Types of Flux

There are various types of fluxes available.

___

Caution

Some types of flux should never be used on a circuit board because they corrode the board and lead parts if the flux is not removed immediately. Acid- or zinc-based fluxes should not be used on a circuit board. Fully activated rosin flux, known as RA, also is not recommended for use on a circuit board.

___

The acceptable types of rosin flux include the pure rosin and the mildly activated rosin (R or RMA). This latter flux is in common use today, with some inroads being made by so-called low residue and no clean fluxes. It has been found that some residues left behind from flux becomes water absorbent and should be removed within a maximum of thirty minutes after the connection has been made.

RA flux is acceptable for use in tinning bus wire or component leads, but should not be used on a circuit board or even kept in the same room, in case it gets picked up and used by mistake. Activators will degrade the board and cause problems that would otherwise not occur.

Most boards operate in an enclosed environment where there are considerable heat, moisture (relative humidity), airborne bits and pieces or dirt. The environment softens the flux left on the board, turning it into a gooey, particle-attracting, water-absorbing blob of useless material. This mess will become conductive as it absorbs moisture, resulting in leakage paths that cause problems in the operation of the equipment.

In late 1992, a new water-soluble flux, developed originally by Hughes Aircraft, received acceptance: final approval for its use was made in early 1993. The flux is made from lemons. This makes it very easy to clean, but the cleaning must be very thorough, or residue may cause corrosion.

WETTING

If the correct flux is properly used, it will greatly assist all aspects of soldering and desoldering. It improves the intermetallic bonding and consequently the solder flow, which is one of the important areas of inspection (see Figure 1–1). Poor wetting is usually the result of poor cleaning procedures or lack of sufficient heat, as shown in Figure 1–2. De-wetting problems relate to the material that is being soldered as a result of the intermetallic compound reaching the surface of the tinned area. If it has reached the surface, the only method of making the area solderable again, is to grind the area down to the copper, flux it and apply a thin coating of solder. The feathering out of the solder on a connection indicates that good wetting has occurred.

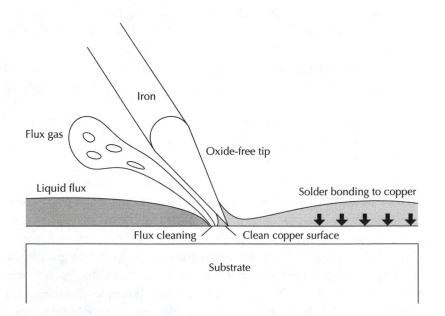

FIGURE 1–1   Interaction of flux, iron, and solder

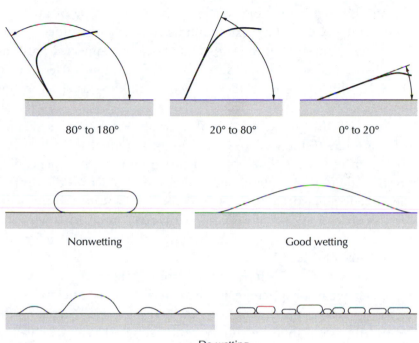

80° to 180°   20° to 80°   0° to 20°

Nonwetting   Good wetting

De-wetting

FIGURE 1–2   Types of wetting action

HEAT, TIME, MASS

The third item in preparing to solder includes three very important factors to be considered. These are the heat to be used, time on the connection, and the mass of the joint. Because not all connections are the same, consideration must be given to the differences in the mass of the joints and adjusting the heat and/or time accordingly. You should not use the same heat and length of time to solder a diode to a small pad as would be needed for soldering a wire onto a terminal. The diode would be damaged, the pad area where the lead is being soldered could be damaged, and the solder will be overheated. An iron that is too cold will result in a mush type of melt and poor wetting action. The maximum time from when a soldering iron comes into contact with the parts that are to be connected until the joint is finished should not exceed two to five seconds. Flex circuits allow very little time for soldering, which could be as short as one second.

One other thing to keep in mind as far as heat is concerned is the oxidation rate of the soldering iron tip. At a normal temperature of 600° F, there is a certain amount of oxidation produced, depending on the time it is left unused and without any solder on the tip (see the preceding Practical Hint). At 700° the rate is nearly ten times the level of oxidation and at 800, approximately hundred times. This oxidation acts as a barrier to the transfer of heat and therefore the proper flow of solder.

Because we are not robots and because people work and react differently to what is happening, it is beneficial for personnel to be able to easily regulate the amount of heat being applied. Changing the heat of the tip of the iron should be the simplest task possible; for example, turning a dial or pushing a button. People should be able to recognize what is actually occurring as compared to what they feel is going on. Experienced solderers have a genuine knowledge of what happened at a particular joint. They know by observing what has happened and can judge whether a joint will be reliable or will break down in a short period of time. To inspect solder connections, a 10X stereo microscope should be made available for managers and supervisors. They should also be trained to know what they should see during inspection.

SOLDER TYPES

The fourth point in preparing to solder is to consider the type of solder to be used. Most companies and technicians use 60/40 solder.

There is nothing wrong with 60/40 solder, but there is a better one—63/37 or eutectic solder. The difference is shown in the graph in Figure 1-3. Note that for the 60/40 solder, there is a time when the solder is neither liquid nor solid. It is in a plastic state during this period. It is very important that there be absolutely no vibration or movement of the connection when the solder is going through this plastic region, otherwise a disturbed joint will be the result. The 63/37 solder has no plastic period and reduces the possibility of a disturbed connection.

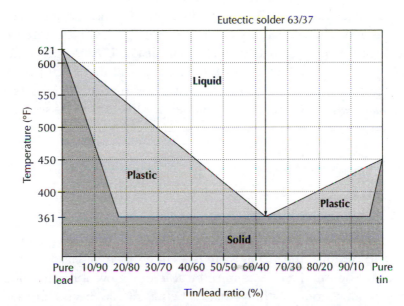

FIGURE 1-3   **Graph comparing types of solder and their plastic ranges. Note that eutectic solder 63/67 does not have a plastic range.**

HEAT SINKING

Heat sinking is a method used to prevent the overheating of components, wires, or circuit boards. It usually is a small metal clip or clamp which is attached to the area between where the solder connection will be made and the item to be protected. The use of a heat sink for soldering some components is not required if the proper technique for soldering is followed to the letter. However, if the proper soldering technique is not followed, heat sinking becomes an alternative, but not a good one. Heat sinking is used mostly by persons who are unaware of the proper procedures. Compensation has to be made for the additional mass of the heat sink by increasing the heat and possibly the length of time. If this concerns a diode in a double-sided board in a plated through hole with a small pad area on each side, the chances of lifting a pad become greater by the millisecond. The heat sink would have to be placed on the top of the circuit board attached to one of the leads. Because the solder and iron are on the bottom of the board, it will be difficult to get the solder to flow through the hole and wet onto the component side pad—which is what should happen for the lead to be soldered correctly.

CLEANERS AND/OR FLUX REMOVERS

Item five in preparing to solder involves the selection of a good chemical cleaner. When it comes to the cleaner to be used for removing flux and cleaning in general, there is a wide variety of cleaners from which to choose. The cleaner must be able to remove ionic and non-ionic residue from anything that is being, or has been, soldered. Check the contents of the cleaner, then check the Material Safety Data Sheet (MSDS) for information on the various chemicals involved. See if anything in the cleaner is carcinogenic (cancer-causing). Even if carcinogens are present in only small quantities, you should try something else. Isopropyl alcohol (IPA), also known as isopropanol, is a decent cleaner; but there are others that contain blends of alcohol that are even better. The key is to find a cleaner that will not harm your work or—more importantly—yourself.

NOTES

CHAPTER 2 Soldering Techniques

Objectives

After studying this chapter, you should be able to:

- Describe the proper soldering technique.

- Describe the visual characteristics of the following joints: overheated, cold, fractured.

- Explain the terms *nonwetting, de-wetting*.

- Describe pinholes and voids.

- List the seven characteristics of a good solder connection.

TECHNIQUE FOR SOLDERING

Various techniques have been tried ever since soldering was first used in electronics. The old saying "the bigger the blob, the better the job" can no longer be accepted. What was considered too meticulous and fussy is now the standard. Soldering can no longer be taken for granted. It is an art, and there are very few gifted painters. In 1989 a person who had received two weeks of formal soldering training in 1981 said, "Soldering sure has changed in eight years!" This comment underscores the need for training from knowledgeable instructors who keep up-to-date with soldering techniques.

The usual soldering technique is as follows: First apply the solder to the tip of the iron, then apply the iron to the area to be soldered, as in Figure 2–1. If flux is not put onto the lead and pad area first, the purpose of the flux in the wire solder is defeated. The flux dissipates

RIGHT

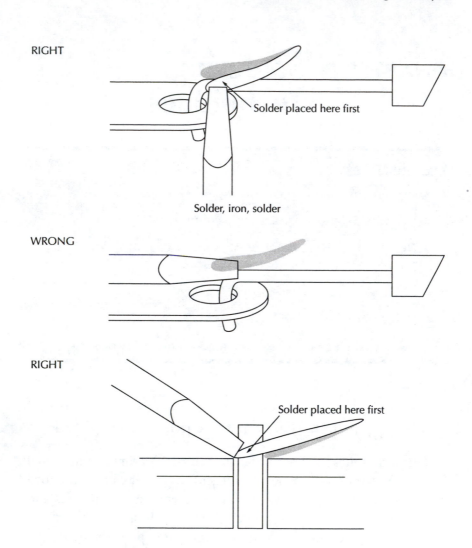

Solder placed here first

Solder, iron, solder

WRONG

RIGHT

Solder placed here first

FIGURE 2–1   Right and wrong methods of soldering

over the iron tip and turns into carbon pieces rather than going onto the lead and pad to remove the oxides. So much for a clean, oxide-free surface; so much for the wetting action; and so much for a good, reliable, problem-free connection.

Solder Application

There are a few exceptions, but the following is a tried and proven technique. Believe it or not, it has been known for decades.

1. **Before the iron is applied,** solder of the proper size is placed beside the lead or wire and on the pad area or terminal to be soldered. A clean iron is applied, and no pressure is exerted on the area being soldered. Only contact with *both* surfaces is required.

2. The proper iron tip—clean, oxide-free, and heated to the correct temperature—is brought to where the solder has been placed, commonly referred to as the "point of maximum thermal mass." As soon as the hot iron touches the solder, the solder melts, permitting the flux in the wire solder to clean off the surface, as well as creating a solder or heat bridge that heats up the joint area very quickly.

3. The wire solder is now moved to the opposite side of the lead or wire, and the proper amount of solder needed to complete the connection is added, Figure 2–2. In either case the exposed copper end of the lead or wire must be sealed by solder to prevent oxidation of the copper, which occurs very rapidly. How do you know if you have the right amount? If the solder is concave and has an angle of wetting between 0° and 20°, it could be a good connection. (See "Characteristics of a Good Connection" at the end of this chapter.)

4. For double-sided and multilayered boards this is the required technique, to ensure that the solder has gone through the board onto the component side and wetted the appropriate area on that side. If this technique is not used, the chances of the solder flowing to the component side, **without excessive heat being applied,** are from very poor to none. Solder should only be applied to the solder side. The solder fillet on the component side of double-sided or multi-layered boards is never applied on the component side.

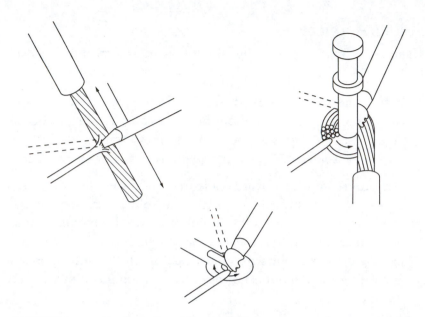

FIGURE 2–2   Preferred solder techniques. In each case, solder is placed first (dotted lines). Then iron is applied so that the melted solder provides a solder bridge for proper heat transfer. Solder is then applied as shown to complete the connection.

Some technicians might have learned that the best soldering method is to apply the iron to the item being soldered and then to add the solder. This method can reduce your chances of making a good connection in two ways:

1. When heat is applied to any metal, the metal oxidizes at a very rapid rate. The higher the heat, the faster the oxidation. This oxidation creates an insulating barrier that will not allow the solder to flow easily onto the surfaces being soldered, thereby preventing the good wetting action needed.

2. Flux in the wire cored solder *should* remove the surface gasses and oxides from the surfaces being soldered. If the solder is

applied after the iron, the overheated flux becomes small pieces of carbon-type material that sit on the soldering iron. The flux never gets to do what it is supposed to do. Worse still, it flows into the connection area, causing the joint to become contaminated—a poor connection.

By applying the solder before the iron, you make proper use of the available flux and form a heat or solder bridge. This technique heats up the surface faster and allows you to complete the job properly in the shortest possible time. As you reduce the amount of time needed to do the job, you also reduce the probability of board damage due to excessive heat and time.

Amount of Solder

When soldering a joint, it is not how much solder is added but the technique used to make the joint. Very little solder is needed in most cases. Usually about one-half to one-third of what is usually considered necessary is all the solder needed. The larger the blob of solder, the more difficult it is to determine if proper wetting of the soldered surfaces has taken place.

Reflow Soldering

A second method of soldering is referred to as reflow soldering. This is normally used where plated through holes are not involved, such as the installation of surface mount items or repairing circuit board traces. The technique is relatively simple and will be covered in Chapter 6.

UNRELIABLE SOLDER JOINTS

Some examples of poor and therefore unreliable connections to watch out for are:

- Overheated—de-wetting; lumps; dull; crystalline-like; looks as though sand has been thrown into the joint.

- Cold—poor wetting, if any; dull; grey.

- Fractured—poor wetting; stretch marks between the pad and lead.

- Non-wetting—solder balled up around the joint.

- Excessive solder—lead or wire contour is not visible and the shape of the solder is convex.

- Insufficient solder—hole is not covered; copper end is not sealed; it is not as wide as the wire or lead.

- De-wetting—usually excessive heat; solder balls up. This also occurs if an intermetallic compound is involved.

    Other defects to watch for include:

- Pinholes or voids—from dust, dirt, flux gas, improper heat, or other

contamination.

- Lumps and large holes—improper presoldering cleaning and out-gassing from flux gas.

- Damaged wire insulation—excess heat and/or wicking of solder under the wire insulation.

CHARACTERISTICS OF A GOOD CONNECTION

A good solder joint has very few things to look for compared to a poor one. A good joint shows the following characteristics:

- Smooth

- Bright

- Shiny

- Clean

- Concave solder fillet

- Good wetting

- The end of the wire or lead covered with solder

"Do it once and do it right" should become the standard for soldering personnel. They should be very familiar with this and be able to apply it to their work. Joints that are heated and re-heated to make them "look" right, become more unreliable with each heat application. Doing the job right the first time eliminates the need for costly rework, equipment returned by the customer, the purchase of rework and test equipment, and the hiring of someone to do the work. Consider these costs and the goodwill of your most important people—your customers.

NOTES

CHAPTER 3   Stripping and Tinning Stranded Wires

Objectives

After studying this chapter, you should be able to:

- Strip 20 and 22 AWG PVC (plastic) and Teflon wire using mechanical and thermal strippers without damage to conductor or insulation.

- Describe the type of mechanical strippers that should be used.

- Describe the two methods for tinning wires.

- Describe *insulation gap*.

- Tin 20 and 22 AWG wires to meet the requirements for a properly tinned wire.

- Explain *wicking*.

- Explain *birdcaging* and why it happens.

STRIPPING AND TINNING METHODS

There are three methods used for the stripping and tinning of various sizes of wire:

1. Chemical

2. Mechanical

3. Thermal

Chemical Stripping

Chemical stripping is used mostly for smaller gauges of wire with very thin insulative coatings. A solder pot can be used to not only strip but to tin the wire in one operation. The problem with this is that because of the contamination caused by the insulative material, the pot must be tested and cleaned out periodically. The solder pot should be capable of maintaining a temperature between 500° and 525° F, plus or minus 10° F.

Mechanical Stripping

Mechanical strippers, Figure 3–1, are used for a variety of wire sizes. The idea is to have a stripper with a fixed die for each size of wire so that it will not nick the wire. The cutting blades should be round to fit the round wire. When used properly, you separate the insulating material a short distance so that the final removal can

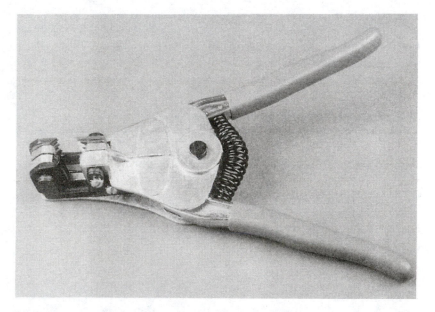

FIGURE 3–1   **Fixed die mechanical strippers.** *Courtesy of Ideal Industries (Canada) Inc.*

be done by hand. This allows the twist of the wire to be maintained as the insulation is removed. Knives or strippers with square cutting edges and similar tools will nick, scrape, and gouge the wire.

Thermal Stripping

Thermal strippers are recommended for wires of smaller gauges. They are usually the tweezer-type with two heated elements that are applied at the point where the insulation is to be stripped. The insulation melts, but is *not burned*, when the heat is applied. The insulation is separated manually, as with the mechanical stripper, the piece is then removed, turning it clockwise to maintain the twist of the wire.

Caution

When thermal stripping wire, PVC does not pose a problem, but Teflon-coated wire poses a very serious problem. Teflon that is burned gives off a very toxic gas, which could have some very serious consequences. If thermal strippers are to be used, the temperature that was used to melt the PVC wire should be tried for the Teflon. Then in very small increments gradually increase the temperature until the Teflon melts. *Do not burn it.* Once the setting has been established, the stripper can then be used without too much concern for successive stripping operations.

A good stripping job will result in no damage to the conductor or insulation. On the other hand, a poor stripping job will leave nicks, cuts, scrapes, broken strands, flared conductors, insulation on the wire, and ragged insulation, as shown in Figure 3–2.

TINNING WIRES

Tinning the wire is not as easy as it sounds or may appear. When done properly, the wire is tinned to within approximately one conductor wire width from the insulation, as shown in Figure 3–3. No

ACCEPTABLE

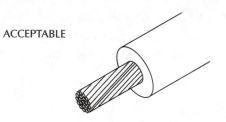

NOT ACCEPTABLE

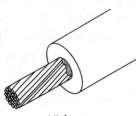

Nick

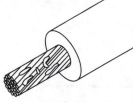

Insulation not completely
stripped away

Broken strand

Maximum surface deviation
is 0.032" (0.79 mm)

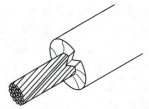

Surface deviation
is greater than
0.032" (0.79 mm)

FIGURE 3–2   Wire stripping

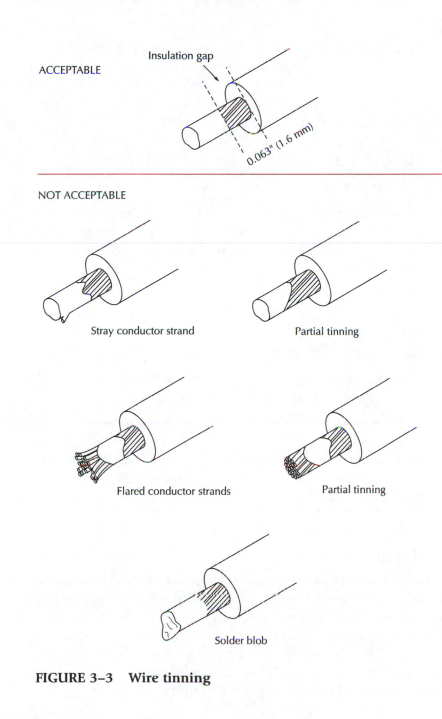

FIGURE 3–3   Wire tinning

solder should go underneath the insulation when the tinning is finished. Solder under the insulation is referred to as *wicking*. Sufficient tinning is needed to prevent the strands of wire from spreading apart when it is being formed around a terminal. The spreading or separation of the strands is called *birdcaging*.

Wicking

Wicking is undesirable for a few reasons.

1. The weakest point of the wire is where the tinning ends due to the solder making the strands into one solid wire. The wire bends a few times where the solid portion and the individual strands meet, and they break. This may restrict current flow.

2. Solder flowing up the wire and under the insulation is pushing flux ahead of it. The chemicals combined with the heat of the molten solder will quickly assist in breaking down a number of different types of insulation.

3. Due to deterioration the insulation will subsequently break off. The bared wire can short out a nearby wire, lead, etc.

Hint

Wicking of the wire is determined by slowly bending the wire. Wherever the wire first bends easily is where the solder flowed to or wicked under the insulation.

Birdcaging

Birdcaging occurs when the strands of wire are wrapped around a terminal and they spread apart, becoming individual strands with no strength and making an extremely weak connection. It is the sign of a very poor tinning job. Birdcaging is undesirable for one basic reason. Tinning a wire is done to make one solid wire out of a number of strands for strength, much like the bundles of cables (rather than individual cables) used for a suspension bridge.

Solder Pot Wire Tinning

Adjustment to time and/or temperature will be needed for different sizes of wires. Even the difference between 20- and 22-gauge wires requires changes to be made. In every case, liquid flux (RMA) is applied to the stripped wire. One method is the use of a solder pot. The fluxed wire is inserted into the molten solder close to the point where the insulation has been stripped. It is then moved over to one side and removed. The solder should come to within one conductor width of the insulation. This is referred to as the *insulation gap*. The wire is cleaned to remove the flux so that it does not become a problem when the wire is installed onto a terminal or into a circuit board.

Soldering Iron Wire Tinning

The second method involves using a soldering iron and the appropriately sized flux cored solder. The wire is held firmly in a downward position, and the solder is placed underneath the wire at the center point of the stripped portion of the wire. The soldering iron is applied at the same point, and when the solder melts, contact is made with the wire by the soldering iron tip. The solder is now placed on top of the wire directly over the iron tip. Both the solder and the iron are moved upwards towards the insulation then down and off the end of the wire. (See Figure 2–2.) Solder likes to flow toward the heat. By having the solder situated on the side opposite to the iron, the heat will draw the solder through the wire, ensuring that the internal strands of wire are also tinned and the inside spaces filled. Solder is added the entire time so that everything is tinned up all the way up and down the wire. A large blob of solder will now be on the iron tip. Once again the solder should stop one conductor width from the insulation (insulation gap). The wire is then cleaned.

ANTIWICKING TWEEZERS

Some people may prefer to use antiwicking tweezers for the tinning of wires if they are unable to eliminate wicking from taking place. If these are used, compensation must be made for the added heat sinking of the tweezers, whether using the solder pot or the soldering iron. Each antiwicking tweezer is a specific wire gauge size. If a number of different size wires are used, each wire needs its own tweezer. Care must be taken to prevent these tweezers from scraping off the tinning already on the wire strands, thereby exposing basis metal or copper. Because of the very fast rate of oxidation/deterioration of copper, every effort should be made to ensure that no copper is showing after the job is completed.

CHAPTER 4 Installing and Soldering Tinned Wires

Objectives

After studying this chapter, you should be able to:

- Describe *insulation clearance,* minimum and maximum.

- Install a wire onto a turret terminal, a cup terminal, and a pierced tab terminal in accordance with established criteria.

TURRET TERMINAL

The turret terminal is one of the most common terminals found on circuit boards. This is the one usually found with the wire wrapped around it two or three times, which makes removal difficult. If the terminal is new, pre-tinning is suggested to ensure solderability. After this operation, remove the excess solder and the resulting flux residue before proceeding. Place a properly tinned wire beside the terminal post and hold it in place with a wooden stick or plastic soldering aid, nothing metallic. Use a pair of needle-nosed pliers to grasp the free end of the wire and pull it around the post, as shown in Figure 4–1. Insulation clearance is the distance from the **base** of the terminal to the point where the insulation stops. The maximum distance should be less than two insulated wire diameters with the minimum being visible clearance between the solder and insulation. (See Figure 4–2.)

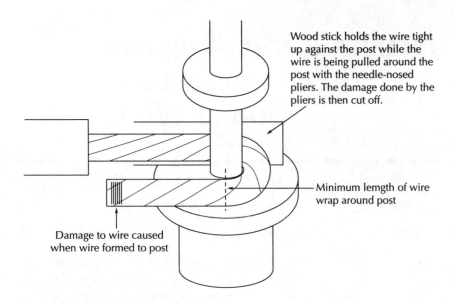

Wood stick holds the wire tight up against the post while the wire is being pulled around the post with the needle-nosed pliers. The damage done by the pliers is then cut off.

Minimum lemgth of wire wrap around post

Damage to wire caused when wire formed to post

FIGURE 4–1   Turret terminal wire wrap

Wire Installation

The wire is cut at 180°. If a 270° wrap is wanted, cut a short distance longer and complete the 270° wrap with the wooden stick or plastic soldering aid. Ensure that the wire is down flat on the base so that if there is any tension on the wire, the post will not bend and cause other problems.

If a second wire is to be installed, it should be installed on the flange above the first wire. If the wires are different sizes, the larger wire goes on the bottom and it dictates the maximum insulation clearance from the base of the terminal. A third wire would be added above the first wire.

Soldering Procedure

Place the proper size solder on the side opposite to where the wire enters onto the terminal. With a clean soldering iron tip of the correct size and temperature, apply the tip at the location where the solder

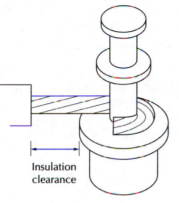

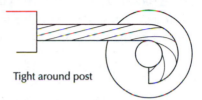
Tight around post

Insulation clearance

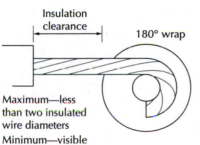

Insulation clearance
180° wrap

Maximum—less than two insulated wire diameters
Minimum—visible clearance between the insulation and final solder joint

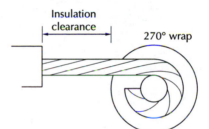

Insulation clearance
270° wrap

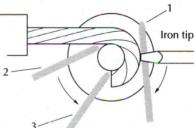

1
Iron tip

2

3

Place solder at point '1.' Apply iron and watch for solder flow. Move solder to '2,' then wipe solder across the cut end of the wire at '3.'

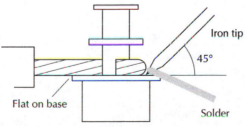

Iron tip
45°

Flat on base

Solder

FIGURE 4–2   Turret terminal

was placed. (See Figure 4–2.) After the solder melts, make sure the iron tip is in touch with the wire and terminal base. Add a small amount of solder where the wire enters onto the terminal, then move the solder to cover the cut end of the wire (the exposed copper). Remove the solder and the iron, with the solder removed slightly ahead of the iron.

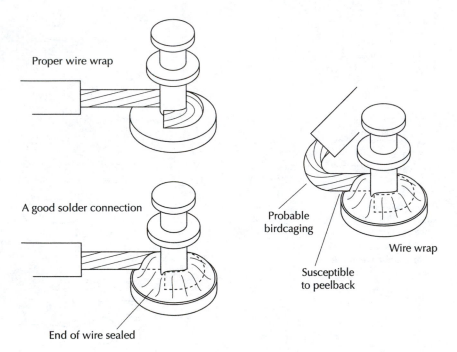

FIGURE 4–3   Wrap, lay, and solder for installing a wire onto a turret terminal

Inspection

Thoroughly clean the connection and check to see that it meets the following criteria:

- Wicking—There should be none if PVC wire is being used. A maximum of 3 mm (0.120") for Teflon or similar type of heat-resistant wire is acceptable, provided the wire is not immediately bent on leaving the soldered connection.

- Good wetting from the base to the wire to the post.

- Proper wetting angle.

- Smooth solder—no holes, pits, or lumps.

- Shiny and bright after cleaning.

- No birdcaging as a result of poor tinning.

- Wire wrapped in proper direction following natural curve of the wire (see Figure 4–3).

- Wire installed and soldered in the proper location.

- Wire contour around the post is still visible.

- Insulation clearance is less than two insulated wire diameters to a minimum of visible clearance between the insulation and the solder joint.

PIN/CUP TERMINAL

Pin/cup terminals are the same design as used for a variety of connectors. They vary considerably in size. The process for soldering, however, is the same. For plugs with numerous pins, the use of a resistance tweezer soldering tool is the easiest method because of the closeness of the pins. This tool goes in cold, not like a soldering iron. The pin/cup completes the circuit for this tool. Heat is applied after good contact is made with the electrodes. (See Figure 4–4.)

Gold Plated Cups

For gold plated cups a slightly different procedure is necessary. Soldering gold with the normal solder that contains tin produces a very unreliable connection. Tiny pores occur at the connection point, and the gold keeps leaching into the tin portion of the solder over a relatively short period of time (days versus years), resulting in a useless, unreliable, brittle solder joint. The gold contamination must

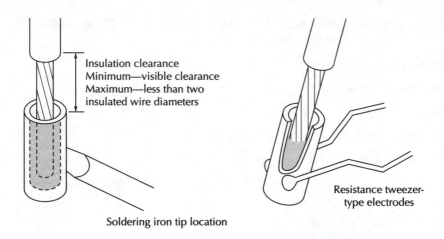

Insulation clearance
Minimum—visible clearance
Maximum—less than two
insulated wire diameters

Soldering iron tip location

Resistance tweezer-
type electrodes

FIGURE 4–4   Cup/pin connector

therefore be removed. This is done simply by filling the gold cup
with solder, then removing the now contaminated solder, leaving a
thin layer of tinning in the cup. The procedure for installing a wire
into one of these terminals is as follows.

Trimming the Wire

Place the wire to be soldered into the cup and establish the depth of
the cup. Remove the wire and trim it to a length that will allow an
insulation clearance of less than two wire diameters between the top
of the cup and the insulation. Insert flux cored wire solder slightly
smaller than the cup into the cup and cut it off flush with the top.

Tinning the Cup

Prepare some solder wick by adding some liquid flux to the braid.
Apply a pristine clean solder tip to the back of the cup and below the
top edge. If a resistance tweezer-type soldering tool is used, it is placed
on either side and below the bottom of the cup opening. When the
solder melts, wait until the flux has bubbled to the surface. as seen in
Figure 4–5. Insert the prepared solder wick into the molten contami-

Wire core solder in cup.

Solder melted to tin cup and flux left on top. Solder then removed with fluxed braid.

Wire inserted to bottom of cup. Solder wets onto wire at front and back of cup.

FIGURE 4–5   Tinning of cup and installation of wire in a cup terminal

nated solder. Remove all of the solder in the cup. Clean the cup of all traces of flux and cleaner. Insert more solder of the same type used to do the tinning into the cup and cut it off at the top of the cup.

Installing the Wire

Again apply your clean soldering iron or resistance tweezer as when tinning the cup. When the solder melts, allow all of the flux and flux gas to come to the top. Insert the pre-tinned wire slowly to allow the tinning on the wire and the solder in the cup to mix together. Make sure the wire goes down to the bottom of the cup. Move the wire to the back of the cup, then slightly forward, then to the back of the cup again and hold it in place. This minor movement in the cup while the heat is applied allows a solder fillet to be formed between the back portion of the cup and the wire. This is the only other item that should be checked when inspecting this type of terminal.

Inspection

No solder, or extremely little, should be left on the back of the cup when the soldering iron is used. No solder should have spilled out of the cup opening onto the outside of the terminal. Check for the same items as when inspecting the turret terminal.

PIERCED TAB TERMINALS

The pierced tab terminal is another very common terminal on circuit boards. When it comes to the soldering of this terminal, it is not necessary (as some think) to fill in the hole with solder. Most tab terminals can have three wires installed, depending on the size of the wire, with three separate and distinct solder joints. If a wire has to be removed, it can be removed and another reinstalled without disturbing the other wire(s). The wire is installed and soldered as follows:

Wire Installation

The wire is placed through the tab opening, leaving an insulation clearance above the top or side of the tab. The wire is bent using a spudger or wood stick and folding it over the tab, as in Figure 4–6. The wire is then cut off flush with the top or side of the tab.

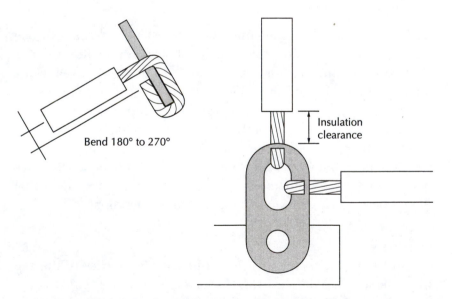

Bend 180° to 270°

Insulation clearance

FIGURE 4–6   Pierced or tab terminal wire installation

Soldering Procedure

The wire is soldered by placing a small size solder (0.4 mm) under the tab and wire in the hole. Apply the iron at this same location. When the solder melts, move the solder to the top of the tab, and apply the solder for the entry fillet, and cover the cut end of the wire. Clean thoroughly and inspect.

BENDING A WIRE

In each case where a wire is installed into or onto a terminal, the wire should be allowed to leave the terminal in a natural curve. If it has to be bent at all, it should definitely not be bent at the solder connection due to the unnecessary stress being placed on the strands of the wire.

NOTES

CHAPTER 5 Components— Through Hole Mount (THM)

Objectives

After studying this chapter, you should be able to:

- Describe why a resistor should be installed above the level of the board surface.

- List the type of terminations normally used for making repairs.

- Name the portion of a button-type ceramic capacitor that is to be kept free from solder and should not be damaged.

- Describe what must be done to transistor leads if they do not fit directly into the mounting holes.

- Describe how a DIP should be soldered into a board— including the number of leads that can be partially clinched (to what degree), when the leads should be trimmed (if needed), and why leads are not soldered consecutively.

- Given a half-watt resistor, a button-type ceramic capacitor, transistor with in-line leads, and a DIP, install each of these items on the board, terminate the leads correctly, and solder them into place. Using a 10X stereo microscope, apply the seven features of a good solder joint as the criteria for inspecting your work.

LEAD SOLDERABILITY

Checking components to ensure solderability may seem redundant or even ridiculous to some, but that depends on the company involved. If the company does not care about its product, this checking will not be done. But if they want to gain a good reputation, it will be done. A random sample of the components to be used for testing can help ensure that the manufactured product will be a good one.

There is a wide variety of components used on circuit boards. This chapter will deal with a few basic components. If the techniques are followed for others, a proper job will be done.

The assembly portion in the automated manufacture of a circuit board is of major importance. The same holds true for hand assemblies. This is what will be discussed over the next few pages as it relates to resistors, capacitors, transistors, and dual in-line packages (DIPs).

CIRCUIT BOARDS

The components just mentioned can be installed in a variety of types and sizes of circuit boards. The majority of boards are made of epoxy fiberglass and variations thereof. Cheaper boards are made of paper or linen impregnated with a phenolic resin. While relatively heat resistant, they have problems with moisture. More-expensive boards are made of ceramic. We will deal with the most common—fiberglass epoxy.

Boards start out with copper covering the entire surface—one side for single-sided boards and both sides for double-sided boards. The copper is either one ounce (0.014"/0.35 mm) or two ounces (0.028"/0.71 mm) thick. The reference to the weight and thickness relates to the spreading of one or two ounces over one square foot of circuit board material. The result is the thickness of the copper indicated above. This is an industry standard. The holes are drilled, and the board is etched to leave the final design of the traces and pads on the board. If the holes are to be plated, the board is put in a plating solution that includes palladium. This is required to deposit an extremely thin layer of copper on the freshly drilled-out holes. The board is then plated with tinning, which doubles the original thickness of the metal on the board and adds to the metal in the holes.

Where the boards have not been properly plated, which is fairly rare today, it is necessary to flux the area and tin the track/trace/run or pad(s) with a very thin layer of solder. This is done by placing flux on the track or pad; when the iron melts the solder, the iron then moves along the untinned area until it is covered. A thin even coating will be the result. The used flux is then cleaned from the board.

RESISTORS

Properly installed resistors will usually last the life of the equipment. The problem is that in many cases this is never considered or is discarded as superfluous. A resistor of less than one watt is normally mounted flush to the surface of the circuit board. If the resistor is one watt or more, it should be mounted above the board or in a heat sink clip to dissipate heat and prevent burn damage to the board.

Lead Bending

The leads should extend straight out from the body so as not to cause unnecessary stress or to break the seal where the lead and body join. The leads are formed to permit them to go down into the circuit board at a 90° angle to the body and centered in the holes. The same bending parameters also apply to vertically mounted resistors. A vertical type of installation should not be mounted flush on the board for a few reasons.

1. It will not permit air flow for vacuum desoldering, which is very necessary

2. It will not allow the solder to flow through to the component side of a double-sided or multilayered board.

3. Depending on the wattage of the resistor, the heat from the resistor could burn the board.

4. It will not provide any sort of vibration or stress relief to the component.

5. Flux and flux gas will also become entrapped in the hole.

Use the correct tool to form the leads and there is no problem. Needle-nosed pliers are not the right tool but round-nosed pliers would be good.

Lead Terminations

The termination of the lead can be straight through or full clinch for normal manual soldering operations involving repairs. Semiclinch is usually used by manufacturers. (See Figure 5–1.) For a straight through lead, the lead should be cut off between 0.5 mm (0.020") and 1.5 mm (0.060"), as in Figure 5–2. *All the lead terminations—except DIPs—should be cut before soldering*. The easiest method to determine an appropriate length in most cases is simply to use another resistor lead. Leads vary in width from 0.45 mm (0.018") to 1.1 mm (0.043") or slightly more. This provides a quick and easy measurement. After putting the leads through the board, place another lead beside the ones to be trimmed and cut them off.

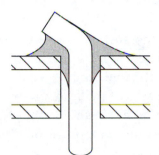
Semiclinch unsupported hole

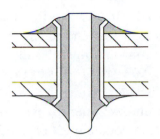
Supported hole with funnellette

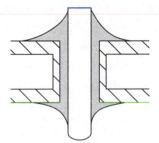
Straight through plated hole

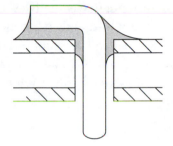
Full clinch in unsupported hole

FIGURE 5–1   Terminations

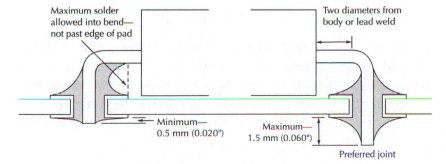

Maximum solder allowed into bend— not past edge of pad

Two diameters from body or lead weld

Minimum— 0.5 mm (0.020")

Maximum— 1.5 mm (0.060")

Preferred joint

FIGURE 5–2   Straight through lead length

Soldering Component Leads

To solder the lead to the board, flux the area and place the solder beside the lead and the pad to be soldered. Make sure contact is definitely made with both surfaces. Apply the soldering iron at the same spot. Move the solder to the opposite side of the lead and pad and add whatever is necessary to cover the hole (single-sided) or fill the hole and wet onto the pad and lead on the component side (plated through double-sided or multilayered).

Soldering should only be done on the solder side, not on the component side of plated through hole boards. The solder joint should extend at least halfway, a maximum of three-fourths the distance up the lead and in a concave form with full pad coverage on the solder side. The component side should also have a concave solder fillet and full pad coverage, as shown in Figure 5–3. Solder should not extend into the bend of the lead past the edge of the pad.

CLINCHED LEAD—PAD ONLY

For a fully clinched lead, if there is only a pad and no track, the lead is trimmed at the edge of the pad.

CLINCHED LEAD—PAD AND TRACK

For a fully clinched lead where there is a track/trace/run attached to the pad, the lead is bent over 90° in the same direction and centered on the track. The leads can be clinched in opposite directions, either away from the component body or underneath the component, but never in the same direction, as shown in Figure 5–4. The lead is

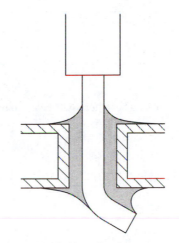

Double-sided board and component

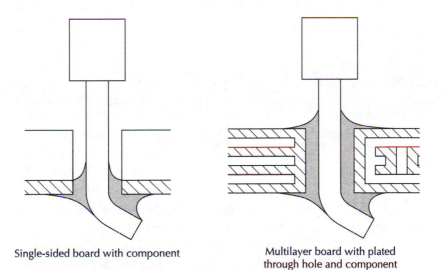

Single-sided board with component

Multilayer board with plated
through hole and component

FIGURE 5–3   Solder joints for specific types of boards

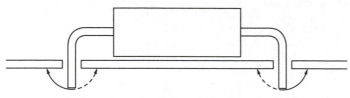

Inward or outward bends for axial leaded components

FIGURE 5–4   Resistor lead forming

trimmed to a length equal to the largest dimension of the pad. If the pad is round, the diameter of the pad is the maximum measurement. If the pad is more square, the length of the lead is determined by the diagonal distance across the pad. (See Figure 5–5.) The minimum length is half these distances.

SOLDERING CLINCHED LEADS

To solder the clinched lead, place the solder at the point where the lead comes out of the hole. Apply the iron at the same location so that when the solder melts, the tip will be in touch with both the pad and lead. After the solder melts, add sufficient solder to cover and/or fill the hole, then move the iron along the lead to the end and remove the iron. Ensure that the solder extends for the full length of the lead along the track, on both sides, and at the end of it.

Removing the Iron

Once the soldering iron is placed on the area to be soldered, it is not moved until it is being removed completely from the finished joint. The solder does the moving around. The iron is removed by

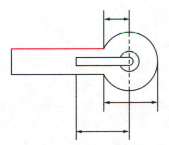

Maximum—one pad diameter
Minimum—1/2 pad diameter

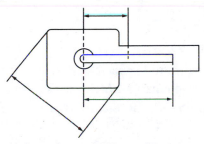

Maximum—one pad largest dimension
Minimum—1/2 largest dimension

FIGURE 5–5   Fully clinched lead

taking it up the lead for a straight through or along the lead for a clinched termination. Moving the iron up or along the lead ensures coverage of the copper at the end, but only if flux has first been applied to the lead.

Note: The above soldering process described here for straight through and clinched leads is the same for all through hole mount components.

CAPACITORS

Capacitors come in a wide variety of shapes and sizes. If the guide-lines are followed, proper assembly will result in reliability. Assembly should include a close examination of the meniscus—the material on the leads below the main body of some capaci-tors—to make sure that it has not been broken or cracked. The meniscus is the seal for the lead and body for this component. Not all capacitors have a meniscus. Any component that has this portion damaged should not be used. Molded box capacitors are easier to use if standoffs are used, but standoffs can not be used for all capacitors.

Button Capacitors

The button-type ceramic capacitor should be installed above the board so that the meniscus on the leads does not end up in the holes. (See Figure 5–6.) The meniscus is the seal for these types of

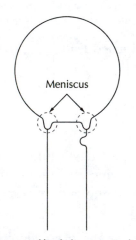

Two different types of leads for ceramic disc capacitors

FIGURE 5–6   Capacitor meniscus

components. For double-sided boards, it is installed sufficiently high above the board so that solder flowing through from the solder side will never come in contact with the meniscus. If the leads have to be formed, this is done so that the meniscus will not be damaged by the bending.

Electrolytics

For large electrolytic capacitors, if the weight on the leads is 1/4 oz or more per lead, the capacitor should be securely fastened to the board.

TRANSISTORS

When leads for transistors are not aligned to the holes, they will have to be formed so that there is no stress at the seal of the body of the transistor. If forming is required, it is done by bending

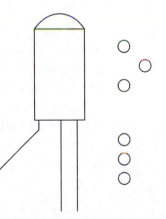

45° bend required if not aligned with pads

FIGURE 5–7   Transistor lead forming

the leads outward a full 45 degrees towards the individual holes where they will be installed. Use a wood stick and place it at the junction of the body and the lead to make the first bend and to keep the seal at the body intact. A pair of round-nosed pliers should now be used to bend the leads a second time so that they will enter straight down into the holes. The transistor is then installed with the newly formed second bend in touch with the pads on the component side of the board. In most cases this is not necessary. But the transistor must be installed above the board to prevent possible heat damage and to allow the solder to come through to the component side.

Standoffs

Where standoffs are used, as shown in Figure 5–8, care should be taken to make sure that the standoff is the right side up and the feet of the standoff are on the board. If the leads are clinched, they are clinched away from the center of the body of the transistor.

DUAL IN-LINE PACKAGE (DIP)

Dual in-line packages (DIPs) are integrated circuit packages that need to be handled with care and caution because of the damage that can be done to them. [See Appendix A on Electrostatic Discharge (ESD).] The only lead forming that should be carried out is after the leads have been inserted into the holes. Up to four leads—two leads on each side—may be partially clinched up to a maximum of 30° from the perpendicular, as shown in Figure 5–9. The body of the component should be parallel to the board surface.

Lead Soldering

Leads should not be soldered consecutively; every other lead should be soldered. This is important because it prevents the buildup of heat

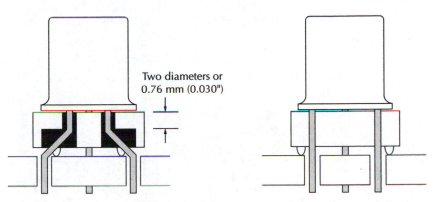

Two diameters or
0.76 mm (0.030")

Parallel mounting of nonaxial leaded components utilizing resilient standoffs

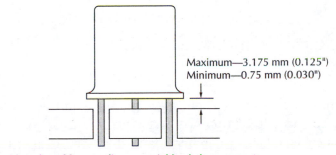

Maximum—3.175 mm (0.125")
Minimum—0.75 mm (0.030")

Mounting of freestanding nonaxial leaded components

FIGURE 5-8   Transistor with and without standoffs

and possible damage, not only to the tiny hairlike internal IC leads, but to the board itself (measling).

Solder Amount and Lead Trimming

The solder should extend no more than halfway up the lead on the component side. The solder fillet on the solder side should be no

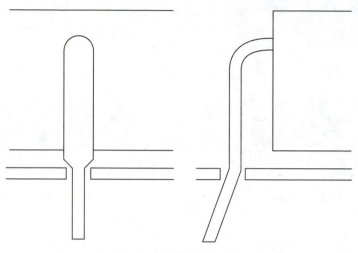

Maximum 30° bend for dip

FIGURE 5–9   Dual in-line package

different than for any other straight through lead connection. If the leads need to be trimmed because of the close proximity to a conductive surface or another board, this is done after the DIP is soldered into the board. This prevents mechanical shock damage to the internal IC leads, which can be caused by the cutters trimming the external leads prior to soldering.

USE OF TAPE

For all of the items discussed in this chapter, when repair rather than assembly is being done, it is difficult to hold the item in place and keep it still while the soldering operation is being performed. To assist in this regard, use some antistatic tape to hold the item. Remove the tape after the job has been completed or when the item has been stabilized. Do not use ordinary plastic tape or masking tape. ESD (electrostatic discharge) can damage what was just installed—or anything else on the board for that matter.

Common sense in respect to heat and time and consideration of the metallic mass of the joint will eliminate such problems as an overheated joint, cold joint, measling, de-wetting, poor wetting, and just about everything that ends up as a poor, unreliable connection.

NOTES

CHAPTER 6 Components— Surface Mount Technology (SMT)

Objectives

After studying this chapter, you should be able to:

- Explain why the change is being made to SMT from THM.

- Describe three methods used to hand solder components onto circuit boards.

- Describe three types of terminations used for surface mount components.

- Explain the following terms: SOIC, SOL-J, PLCC, CLCC, TCE, footprint, coplanarity, and Type II assembly.

- Given a chip resistor, 14 lead SOIC and 20 lead PLCC, install each of them onto a circuit board and solder them in place. Using a 10X stereo microscope, apply the seven features of a good joint as the criteria to inspect your work.

SURFACE MOUNT TECHNOLOGY

Surface mounting is the installation and soldering of the leads and termina-tion areas of devices and components on the surface of one or both sides of a circuit board.

This technology, though looked upon by some as just being a fad, has been with us in different forms since the late 1960s. It is only in the last few years, however, that manufacturing equipment has become sufficiently sophisticated to be able to easily cope with the small components and fine-line leaded devices. People who feel that this is only a fad have obviously been left completely behind the times. Estimates from equipment manufacturers indicate that in 1995 as much as 75 percent of electronic board manufacturing was SMT. Today it is much higher.

Reasons for SMT

There are a number of reasons for the change to SMT, reasons that relate to society in general. People want things to be small, to be lightweight, to be easy to handle and operate, to do anything imaginable, and—of course—to do it fast. SMT allows all of this to happen.

As a result of miniaturization, there is weight reduction, the costs of materials have been reduced, and in some cases, manufacturing costs have been reduced due to the elimination of hole drilling requirements. This is a very costly procedure due to constant drill bit replacement and programming for the machine to perform the job.

Terms Defined

As with any new technology new terms come along to identify that technology. Surface mount technology is no different. The following list presents some terms to become familiar with for the future:

SMC/SMD—Surface mount components/surface mount device. *Component* usually refers to passive items such as resistors, capacitors, etc. Device refers to active items such as SOIC, PLCC, etc.

MELF—Metalized electrode face (or most end up lying on floor). Cylindrical and usually color-coded resistors or solid-colored diodes.

SOIC—Small outline integrated circuit. This smaller version of the DIP has leads that are gull-wing shaped to allow connection to the board surface (see Figure 6–1).

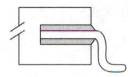

Gull wing or extended lead

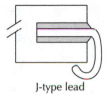

J-type lead

SOT

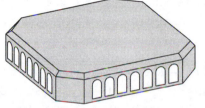

Leadless chip carrier (LCC)

SOIC-8

Narrow body SOIC-14

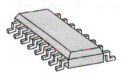

Wide body SOL-16

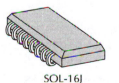

SOL-16J

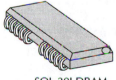

SOL-20J DRAM

FIGURE 6–1   Sample of SMT devices

PLCC—Plastic leaded chip carrier. Normally a four-sided quad package in which an IC is installed with J-type leads extending out from the sides of the package then down and rolled under the body of the device.

SOJ—Small outline J leaded package similar to a DIP but for surface mount applications only.

SOL-16J—Small outline, 16 J leads.

Footprint—The metal pads on the substrate surface intended for mounting specific SMC/SMDs.

SOT/SOD—Small outline transistor/small outline diode. A plastic leaded component in which diodes and transistors are packaged. Gull-wing leads are used.

TCE (CTE)—Thermal coefficient of expansion (coefficient of thermal expansion). The rate of expansion or contraction of a material when its temperature is increased or decreased.

Coplanarity—Each lead of a multileaded item being at the same level or plane. This is extremely important to ensure that all leads are properly soldered.

LCCC—Leadless ceramic chip carrier. A ceramic package with an IC mounted to form a surface mount device. Its termination areas for soldering are built into the ceramic material and do not allow for TCE of the device versus the substrate, especially if the component and board materials are different.

QFP—Quad Flat pack. Four-sided device normally with extended or gull-wing type leads.

TSOP—Thin shrink small outline package. Similar to SOIC but smaller packaging and closer lead spacing (8 to 24 leads) with the leads protruding from the ends of the package.

Type I Assembly—Circuit board with SMC/SMDs on one or both sides.

Type II Assembly—Circuit board with a combination of THM on the top of the board and SMC/SMDs on the top and bottom of the board.

Type III Assembly—Circuit board with THM on the top of the board and passive surface mount on the bottom.

Solder paste/cream—Semi-liquid substance made up of tiny solder balls, flux, solvent and an antislumping agent,

Reflow soldering—The reflowing of the solder paste/cream used to solder SMC/SMDs onto the board surface. The solder paste/cream is heated until it melts and flows, usually using an infrared, convection or vapor phase heating system.

Measling—The damage to a circuit board caused by overheating. Usually shows as small white dots on fiberglass epoxy boards around the overheated area. This is the weave of the fiberglass separating inside the board.

Blind via—Surface mount connection hole where the board is multi-layered and the hole is attached to the surface (top or bottom) and only goes partway through the board.

Buried or Interstitial via—Similar to a blind via, except that it connects internal layers only and is not exposed to top or bottom surfaces.

Ball grid array (BGA)—A large IC carrier with small solder protrusions covering the entire surface of the bottom of the package for attaching to the appropriate pads on the circuit board. Difficult if not impossible to remove/install without the proper equipment.

Specialized Tools and Equipment

As you might imagine, there are many types and varieties of tools to deal with SMT devices and components as there are high-tech companies. They vary from modified soldering iron tips to specialized hand tools with variable heat controls, tweezer-type with variable heat and hot-air systems in hand tools as well as large rework stations.

Testing

The ability of the operator to cope with the different tools dictates how well work is performed. Considering this, why not have the employee test the equipment personally? Most distributors and dealers are more than happy to permit their tools or equipment to be tested, so take advantage of it. Try them all, not just the first one you see. Make sure that the person who will be operating the tool or equipment is going to feel good about using it. (Remember to shop wisely. Many people eventually regret having purchased the cheapest model.)

Find out if training can be provided for new tools and equipment. Perhaps training sessions can be included in the purchase agreement; perhaps someone with the necessary expertise can be hired to provide training sessions for a day or two.

Alignment

Because of the size of IC leads (Figure 6–2), components, and footprint/pad areas, very close attention is required in order to ensure that the components and devices are properly aligned to the footprints. The latest permissible misalignment is in the order of 0.0001", and it is getting less with every new device.

Inspecting SMT

Microscopes are needed for SMT and should be used for final inspection. In some cases microscopes are necessary during the soldering

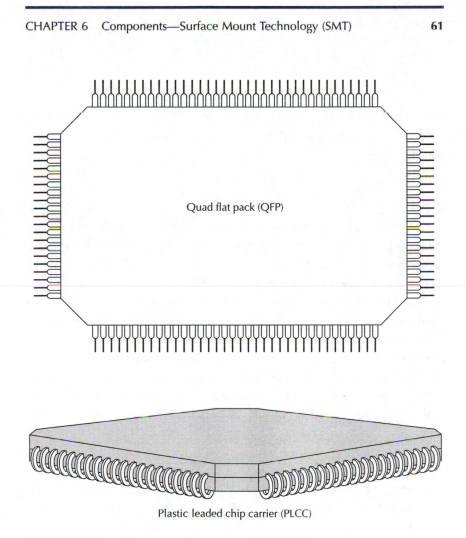

Quad flat pack (QFP)

Plastic leaded chip carrier (PLCC)

FIGURE 6–2   Two popular types of SMT ICs

process. This is the only way to make sure that the item is properly soldered into place and that the necessary wetting action has occurred. A stereo microscope of 7X to 20X is the best for inspecting SMT solder connections. These items are no longer considered a luxury but a necessity.

SOLDERING PROCEDURES—SMT

There are three methods used to hold and solder SMT items onto boards: tinning and reflow, tack and solder, solder paste and reflow.

Tinning and Reflow Method

The tinning and reflow method uses wire-cored solder with RMA flux. Apply a small amount of liquid flux (RMA) to the footprints or pads and flow a small amount of solder onto the pad. The flux residue is cleaned off and new flux added. The item is placed onto the solder and the solder is reflowed to make the connection between the footprint/pad and the lead of the component/device. This method, though used by some for passive components, is not recommended for more than one pad. The reason is that the unsoldered leads or terminations of the various components/devices will not be in the same plane (coplanarity) as the leads that have been soldered. This results in stress to the lead and solder connection. For example, using this method for an eight lead SOIC, the eight footprints are first fluxed, then a small amount of solder is applied. After cleaning and refluxing, heat is applied to the lead and solder of one footprint. When the solder flows, the lead is then forced down onto the footprint with a wooden stick and held down until the heat is removed and the lead cools. This lead is now undergoing stress since the other leads are still sitting on top of the solder above the plane of the first lead. This makes it virtually impossible to properly solder the remaining leads as they would have to be forced down, resulting in tremendous stresses to the device, leads and solder joints.

Tack and Solder Method

A second soldering method is to tack one or more leads to one or more footprints using the small amount of tinning already on the lead and footprint. Coplanarity of the leads is necessary for this to be done properly. If any leads are not at the same plane, tacking will be impossible for those particular leads. Tacking will hold the component/device in place while soldering the lead(s). Tacking is done by applying flux to the surface of the footprint(s), positioning the component, then applying the soldering iron to the footprint only, not to the lead, to melt the tinning on it. The molten solder wets under the lead and on top of the footprint. This small amount of liquid solder will be sufficient to stabilize the component so that it will not move while the remaining leads/terminations are soldered to the other footprints.

The areas that were not tacked are soldered first, then the areas that had been tacked are done. Liquid flux is absolutely necessary to perform this operation successfully.

Solder Paste and Reflow Method

A third soldering method—and probably the most used—is to dispense solder paste/cream onto the footprint area, place the component/device into the paste, and use a hot-air or other type of hand tool to reflow the solder paste and make the connection. Coplanarity applies to this method as well. Some people use the term *solder paste* while others use the term *solder cream*. These are the same. If an adhesive is used to hold the item in place, a person may want to clarify things by using the term *cream* for the solder. Nearly all solder pastes/creams can be used immediately and do not require a curing time, which used to be a requirement when they were first introduced.

Soldering irons are not recommended for use with solder paste. Solder paste requires a gradual increase in heat to dissipate the volatiles in the paste. A solder iron will not do this and will splatter the paste, producing solder balls, which end up everywhere on the board.

SPECIFIC TOOLS

Any one of the three methods of installation just discussed can be used for any SMT item. One method might be far better than another in the hands of a certain individual for one specific operation. Another tool may be required for a different item. When it comes to SMT, one hand tool cannot do the entire job.

When it comes to the amount of solder needed for the proper installation of SMT components or devices, the thickness of the solder between the termination point or lead and the footprint is extremely small—about 0.075 mm (0.003"). This is not the solder fillet itself but the space between the two surfaces that meet on the board.

SOLDER JOINTS

For dual termination components the solder joint should extend from the top of the termination to the edge of the footprint in a concave shape, shown in Figure 6–3.

Equal amounts of solder at each end of the component will eliminate problems of joint or component cracking, due to unequal stresses being placed on two different sized joints during thermal

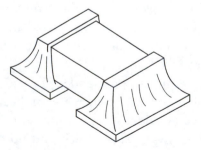

FIGURE 6–3   Solder fillet for surface mount termination

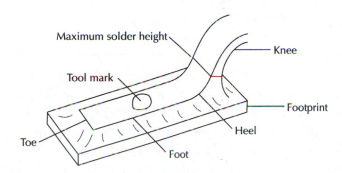

FIGURE 6–4   Gull-wing or flat pack leads

variations. Variations are brought about by temperature changes in the working environment, seasonal changes, air conditioning and the conditions under which the equipment is operating, to name a few. The equal amounts of solder are for inspection purposes to ensure that the solder has properly wetted over the soldered surfaces. Minimum height for the solder should be three-quarters the height of the termination area.

For gull-wing leads, the connections are at the end of the lead (toe) and footprint and at the heel of the lead. The concave fillet at the heel should extend no farther than halfway up the angled or vertical portion of the lead. See Figure 6–4.

For J leads, the connection should have a concave solder fillet all around the lead where it meets the footprint.

THERMAL COEFFICIENT OF EXPANSION

One of the early problems with SMT was the use of ceramic components and devices on noncompatible substrates. This incompatibility is due to the differences in the TCE of the various materials used. For example, ceramic expands at a rate of 6.5 ppm/°C whereas fiberglass epoxy expands at rates of 12 to 16 ppm/°C. So the epoxy material;

expands at two to two and one-half times the rate of ceramic. This is very hard on the solder connections, for instance, where a ceramic capacitor is installed on an epoxy-based board. The solder joint is subjected to some severe stress. If the solder connections are not properly completed, the joints will crack. The application of high-temperature soldering irons to small ceramic capacitors has caused instant destruction or serious damage to the component. Extremely small internal connections are shown in Figure 6–5. To circumvent this concern, equipment is now available that incorporates a "ramp-up" in the temperature so that the sudden heat will not damage the component. The rise time is in the order of $11\,°F/6\,°C$ per second; these are commonly a tweezer-type tool. You can see by this that it will take much longer than usual to solder the component using this equipment.

Another method that would prevent damage to these same capacitors is preheating the component and the circuit board to approxi-

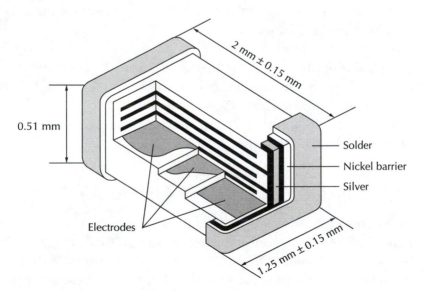

FIGURE 6–5   Surface mount chip capacitor

mately 200°F/93°C. This will reduce the extent of the thermal shock applied to the component or board. One problem with pre-heating a part and board, is the added oxidation produced by the heat and therefore an unwanted barrier to the solder. Liquid flux is a definite requirement.

SOLDERING METHODS—SMT

The following methods are some of the ways in which components/devices can be installed. How they are installed is totally dependent on the hand tools available. Tools can include a soldering iron, hot air handpiece, tweezer-type tool which gradually heats up and has untinnable tips, and a tinnable, continuously heated tweezer, to mention a few. When using solder paste, a soldering iron is not recommended as it will explode the paste and cause solder balls to occur under and around the item being installed, instead of the paste flowing smoothly out over the surfaces of the termination/lead and pad areas. A good tool for reflowing solder paste is a hot air jet which will gradually heat up the paste, allowing the chemicals to be eliminated to prevent the solder balls. Start out with the air jet about 3/4" (2 cm) above the paste and move the handpiece to ensure that all of the paste is being heated up. It will change to a light grey color just before the solder flows. Move the handpiece closer to the paste and keep it there for a second **after** the solder has flowed over all of the surfaces.

Chip Resistors

1. Pre-tin one of the two pads. Add flux, align the resistor and holding the component with a wooden stick or tweezer, reflow the solder on the tinned pad onto one of the terminations. Make sure you have a tip size at least the same width as the chip. After reflowing the solder, bring the tip up, to get the solder to the top of the termination area. After doing this, if you

notice a spike at the top of the solder connection, turn your heat down or do it faster. The solder is following the iron because of the excess heat or time being taken to do the job. Add flux to the opposite end of the resistor, apply the solder, then the iron to make the second connection, again bringing the iron up the termination area. Try to ensure that the same amount of solder is used for each connection.

2. Apply solder paste to both pad areas, align the resistor and use a hot air handpiece with a single or dual nozzle, to **slowly** reflow the solder paste. Solder balls will result from trying to reflow the solder paste too quickly. The other chemicals in the paste need time to do their job and burn off.

Chip Capacitors

These items pose a particular problem due to how they are manufactured. They comprise layers of alumina (ceramic) and metal. When heated, the metal expands faster than the ceramic resulting in internal separation between these materials. The preferred technique is to heat these items up gradually whether with hot air or a tweezer-type tool that will gradually warm up to solder temperature. The item may work initially but will break down due to the continual heating and cooling of the component while in operation.

1. If ceramic capacitors are preheated prior to installation to prevent heat damage, a soldering iron at a low temperature could possibly be used. The problem with preheating the capacitors, is the buildup of oxidation on the termination areas.

2. Use the same method as indicated for Chip Resistors, method 2, discussed previously in this chapter.

3. Apply solder paste and using a tweezer that heats up gradually, pick up the capacitor with the tool and position it over the pad areas. Apply the heat and remove the tweezer after all of the solder

paste has reflowed. With the gradual heating up of the tool, it allows the other chemicals to do their work.

MELF

1. Methods 1 and 2 under Chip Resistors in this chapter can be used for MELFs.

2. Method 3 in Chip Capacitors, the previous section, can also be used for MELFs.

3. Pre-tin both pad areas, add flux, pick up the MELF with the pre-heated tweezers, align the MELF over the tinned pads, and apply the heated tips to the pads concerned. Ensure that the solder melts completely and that it flows up the termination points.

Refer to Figures 6–3 and 6–6 for the proper amount of solder for the above three types of components.

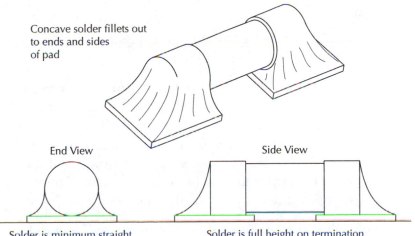

Concave solder fillets out
to ends and sides
of pad

End View

Side View

Solder is minimum straight
down to concave shape

Solder is full height on termination
areas and concave shape. Minimum
is three-fourths the height

FIGURE 6–6   MELF

SOT

1. Pre-tin the pad where the single lead is to be installed. Apply flux, align the SOT on the pad and place the soldering iron at the end of the lead, in contact with both the lead (toe area, see Figure 6–4) and the pad, and reflow the solder. For the other two leads, flux, apply the solder at the toe then the iron. Solder of 0.25 mm (.010") thickness and using the width of the pad area as the length, is sufficient to do the job. No needle point or conical-type tip.

2. Apply a small amount of solder paste to the three pad areas and using a single or dual air jet, reflow the paste. Don't forget— **SLOWLY**.

SOIC

1. Tack two leads, see Tack and Solder Method in this chapter. Apply flux. Then, using the 0.25 mm (.010") solder, solder every other lead down one side using the same technique as for SOT leads. Go back and do the leads that were skipped, and then do the opposite side in the same manner. If you remember the reason for doing it this way for DIPs in through hole soldering, you will understand why the same applies here.

2. Apply flux to all of the SOICs footprints. Hold it in place with a tweezer. Use a single-sided iron tip with a scooped out end and fill it with solder so that it hangs below the flat line of the tip. No solid metal contact is made between the actual iron tip and the lead. Draw the tip along the one side, again **slowly**, and it's done. Do the same thing for the other side. It takes a bit of practice, but it is easy and fast once the technique is perfected.

3. Apply solder paste in one continuous line from the outside edge of the first pad to the outside edge of the last pad on both sides. Using a hot air handpiece with a single nozzle, do one side then the other. If a dual nozzle is used, both sides can be done at the same time.

4. Apply solder paste as in the preceding method 3. Using a hand-piece with a bar-type tip, apply the bar to all of the leads at the same time. Turn on the heat and make sure all of the solder paste reflows.

PLCC/SOJ

1. Tack solder one lead at one corner then the diagonally opposite corner to ensure proper alignment. Because one side is aligned, doesn't mean that all four sides are aligned. Then tack solder the other two corners so that all four corners are tacked and the device is aligned, see Tack and Solder Method. Using the proper-sized solder, place the solder at the junction of the lead and pad and apply the soldering iron tip to flow the solder. See Figure 6–7 for the proper amount of solder for a J lead. Do every other lead on all four areas, then go back and fill in the ones that were skipped.

2. Tack solder one lead on each side at one corner then one lead on each side at the diagonally opposite corner, see Tack and Solder Method. Apply solder paste from the outside edge of the first pad to the outside edge of the last pad for a section of leads. Apply the hot air handpiece **slowly,** for the same reasons as previously indicated. Make sure all of the paste has reflowed and there are no solder balls running around under the device.

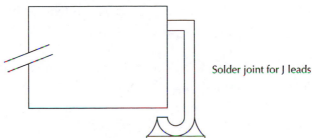

Solder joint for J leads

FIGURE 6–7   J lead termination

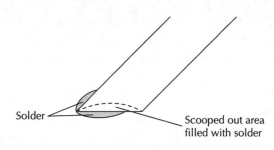

Solder — Scooped out area filled with solder

FIGURE 6–8   Single-sided soldering iron tip

3. Tack solder one lead on each side at one corner then one lead on each side at the diagonally opposite corner, see Tack and Solder Method. Apply flux and using the single-sided soldering iron tip that was mentioned for the SOICs, fill the scooped out portion of the tip as before, but also apply solder to the back side of the tip as shown in Figure 6–8. Apply the iron to the leads so that the back side of the tip makes contact with the lead and the single sided portion of the tip makes contact with the pad, or at approximately a 45° angle. Move the tip along the set of leads, ensuring that the solder is adhering to the leads and pads. In a tight location, it may be necessary to reverse the iron and place the scooped out area with its bulging solder against the lead with the back side of the tip down on the pad. This allows a little easier access where other small components or devices may be concerned.

4. For companies where rework or replacement is required on a regular basis, larger more involved pieces of equipment may be needed to reduce the time factor. These sometimes have a CCD camera for placement of the device in one location and another camera over the reflow area to ensure that all of the paste has reflowed. Solder paste is applied, the device aligned, and then the hot air is turned on to reflow the paste.

QPF

1. Alignment of these items is very critical as the leads are getting extremely close. Tack all eight corner leads, see Tack and Solder

Method. Check the coplanarity of the leads. This is very important for this type of device. Tack solder two leads, see Tack and Solder Method. Using the single-sided tip, use the same method as previously described for the SOICs, for each side. For devices that have a very fine pitch, the leads will act almost like solder wick and tend to draw solder up along the leads. This will result in considerable bridging. The length of time taken to do the job will determine the amount of bridging that may occur. Smaller tips with scooped out tips are also available for fine pitch leads.

2. This method involves the same machine as previously indicated for the PLCC/SOJs, method 4. Solder paste is applied, the device is aligned; then the hot air is turned on and the paste is reflowed.

CERAMIC BOARDS

Ceramic boards should be treated in the same manner as ceramic capacitors. Sudden high heat applied to a ceramic board could result in a crack—in some instances severe enough to render the board useless and costing a considerable amount of money to replace. Ceramic boards are much more expensive than fiberglass epoxy boards.

MULTILAYER BOARDS

Multilayer boards should probably be preheated as well before attempting to remove or install items. Internal heat sinking problems are created by the voltage and ground planes normally comprising the middle layers of the board.

HOT AIR FOR SMT

A hot-air system and/or hand tool are types of equipment that can be very good on boards and components when properly applied, especially for ceramic materials. They allow gradual rise in the temperature, and there is no metal-to-metal contact. The possibility of pressure being applied and lifting a pad is greatly reduced.

CHIP RESISTORS

One other area to be aware of is handling chip resistors. The resistive material is under a protective glass film. This can easily be scratched and the underlying thick film resistance material damaged by a metal tweezer, thereby changing the value of the resistor (see Figure 6–9).

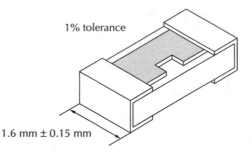

1% tolerance

1.6 mm ± 0.15 mm

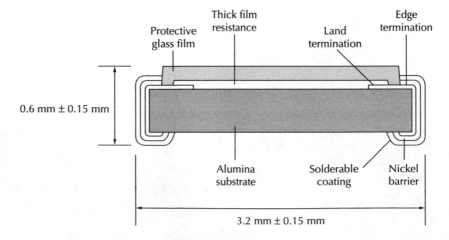

Protective glass film

Thick film resistance

Land termination

Edge termination

0.6 mm ± 0.15 mm

Alumina substrate

Solderable coating

Nickel barrier

3.2 mm ± 0.15 mm

FIGURE 6–9   Surface mount chip resistor

Use tweezers of a softer material or a vacuum pickup to prevent this damage from occurring.

The only answer for people having to deal with SMT is suitable training and practice, and more practice, and after that more practice.

NOTES

CHAPTER 7 Desoldering THM and SMT

Objectives

After studying this chapter, you should be able to:

- Describe a resweat joint and explain why it occurs.

- Describe what has to be done before attempting to desolder a through hole mount component.

- Describe what must be done if solder is not completely removed from a plated through hole.

- Describe the different technique required for the removal of surface mount items and why it is necessary.

- Given a series of three different through hole mount boards, remove five components from each board without causing damage to the boards, pads, or traces.

- Given a surface mount board, remove six items from the surface without damage to the board, footprints, or traces.

DESOLDERING

Desoldering is a topic that no one seems to get too excited about until the time comes when people discover that a lot of damage is being done during this operation.

What happens if a solder connection is not done properly and you have to fix it? What do you need to do? If you want to do it correctly, the connection will need to be desoldered then resoldered properly.

PREPARING TO DESOLDER

What needs to be done before attempting to desolder? The same criteria for soldering also apply to desoldering: clean surface, use of flux, and proper heat and time, depending on the mass of the joint.

POOR DESOLDERING

The following three examples show what happens when uninformed individuals try to solve desoldering problems. These represent actual experiences of the author.

1. A circuit board was brought in because this person "thought he had a problem." On examination, twenty-two areas of damage were found, which included eight lifted runs and fourteen lifted pads. Someone obviously had no idea what had to be done to safely remove the solder from one CPU device. A $2,000 computer board nearly became scrap because someone did not know what to do.

2. A piece of desoldering equipment was sold to a company. A couple of weeks later they brought in a board because the operator was having trouble removing the solder from a row of connectors. Using the purchased piece of equipment, a connector was quickly removed by the author. When the board was taken back to the company by the salesperson, repair people were told to "clean and flux" first. The area where the operator had been trying to desolder proved why the job was not successful: it was extremely dirty. The problem was resolved.

3. A supervisor from a communications company brought in a small pager board because his people were having problems removing a particular "can." While the supervisor watched, it was desoldered without a problem. The same old problem: Old and dirty equipment and lack of knowledge about what to do.

RESOLDERING

For a connection that has just been made but needs to be redone properly, the use of flux is not usually necessary, but there is nothing wrong in using it. The flux will clean the surface during the desoldering process. Don't forget, though, that flux does not remove greases, grime, or other dirt. Flux is for oxides only.

DESOLDERING METHODS

Different methods of desoldering are required, depending on what has to be removed. Of course there are a variety of pieces of equipment to do this job.

Solder Wick or Braid for Through Hole Mount (THM)

For desoldering items on single-sided boards, one of the most common methods—but certainly not the best—is the use of solder wick. This is a copper braid with dried flux included in the braid. Even with the dried flux already in there, the task is better accomplished if liquid flux is applied before the wick is used. The braid is placed on top of the solder joint, and an iron is applied to the braid. A capillary action starts due to the holes in the braid; the molten solder is drawn into the wick. There are different sizes of braid for the various sizes of solder connections.

One of the problems with using solder wick is the pressure applied in attempting to remove the solder. This is natural, except that people apply more pressure than is necessary. Even the weight of an iron is too much, but operators, almost to a person, add additional weight/pressure to the connection. The excessive weight can result in measling, damage from excessive heat, and/or lifted pads or runs from wherever the solder is being removed. This is because the heat and pressure applied have overheated the adhesive holding the pad or run to the board. Damage has been done that could have been avoided if these facts had been known. If wick is used, cut off the used portion. It is useless and will act as a heat sink, resulting in excess heat and time being required for the next solder removal task.

Solder wick is definitely not a recommended method for double-sided or multilayered boards due to the excessive heat needed and

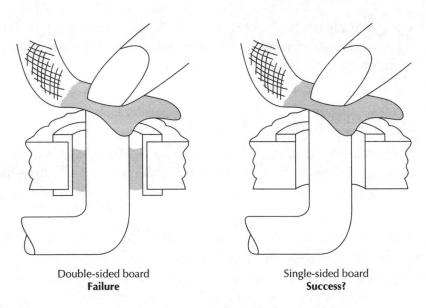

Double-sided board
Failure

Single-sided board
Success?

FIGURE 7–1   Use of solder wick (braid) for desoldering

the fact that solder is left in the hole, as shown in Figure 7–1. This small amount of solder is strong enough that attempts to remove it meet with a very good chance of removing part of the hole. A simple job becomes a major repair, especially on a multilayered board.

HEAT AND PRESSURE PROBLEMS

Did you know that when a pad or lead is heated up, the glue holding the copper onto the board loses 80 percent of its strength? Now you know why it is important to use minimal heat and no pressure.

The pressure/weight problem can be controlled by having the board placed in a holder and turned so that it is perpendicular to the bench surface. This eliminates the natural downward weight of a person's hand, wrist, and arm; it is left up to the person doing the job to apply that infinitesimal amount of pressure needed.

OTHER DESOLDERING TOOLS

The squeeze bulb or spring-loaded solder pullit should not be used for desoldering because of the likelihood that they will affect items sensitive to electrostatic discharge (ESD). The air flowing through the plastic or Teflon tip generates, in most cases, a great deal of static electricity. This is not something you want to use around sensitive boards or devices.

SWEAT JOINT

A resweat joint will occur if either the squeeze bulb or spring-loaded solder pullit is used. The cold tip and cold air make the joint difficult to remove. Because the solder cannot be kept at solder melt tempera-ture by the tool, it is unable to completely remove all of the solder from the plated through hole. When the tool is removed, the lead will make contact with the side of the hole and reattach itself to the molten solder still around because the iron is still there.

Some of these tools have gone from Teflon to ceramic-type tips. Ceramic is a very cold material. Consequently the solder connection has to be overheated to make sure the solder remains molten while it is being removed. You already know what can happen to the board and pad when excessive heat is applied.

MULTIPLE LEAD DESOLDERING

Tools for desoldering all of the leads of a DIP at the same time can be a cause for some concern. All of the leads are not connected to the same mass at every joint. If the tool is applied with sufficient heat to melt all of the leads at once, the heat may be alright where there is only the pad. What about the pad(s) where the voltage/ground plane is attached? It will not be enough for them. So more heat is needed to compensate for their larger mass. What happens now to that one solitary pad sitting there? Overheated adhesive overcures and loses its strength. The result is a damaged lifted pad—and more repairs.

VACUUM DESOLDERING

A good desoldering tool should be able to remove all of the solder in one operation, leaving the hole free and clean of solder without the leads re-sweating to the sides of the holes. These are powered vacuum desoldering units. The board is placed in a holder and turned so that it is at right angles to the bench top. One reason for turning the board sideways is to prevent the desolderer from applying excess downward pressure just from the normal weight of the hand and arm, that would result if the board was placed flat down on the bench surface. A second reason for placing the board in this position is to allow the person doing the desoldering to see when the solder has melted on a double-sided or multilayered board. This tool is placed over the lead with the proper-sized tip, and movement of the tip starts immediately. When the solder becomes molten all the way through the hole, the operator will see the solder moving on the component side. This is when the vacuum is activated and the operation completed as quickly as possible.

Desoldering Technique

When the vacuum pump is activated to remove the solder from a plated through hole, the lead over which the desoldering tip was placed is moved in a circular motion for round leads (Figure 7–2) and back and forth for flat leads. The vacuum removes the solder from around the lead and brings in cool air to lower the temperature of the hole and the lead to below solder melt. This prevents the lead from resweating to the side of the hole and allows the component to be removed easily.

The vacuum should be kept on until after the desoldering tip is removed from the lead. This gives the molten solder time to get up to the collection chamber. If that time is not provided, the solder will end up in the heater, clogging it up and aggravating the operator— the one actually responsible for the problem. Newer units have a time delay, preventing operators from turning off the vacuum too soon.

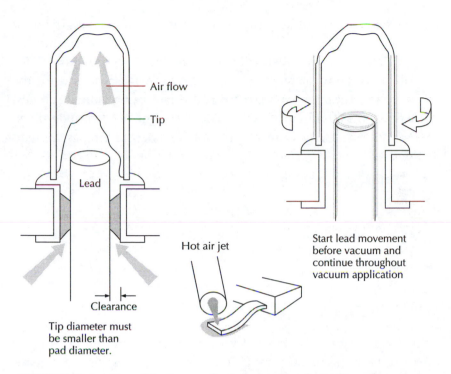

Air flow

Tip

Lead

Clearance

Tip diameter must
be smaller than
pad diameter.

Hot air jet

Start lead movement
before vacuum and
continue throughout
vacuum application

FIGURE 7–2   Desoldering techniques using hot air and vacuum

Old Connections

If desoldering is required of a connection that has been in service for
a long time, the first task is a thorough cleaning of the joint and area
to remove the unwanted dirt. The dirt acts as an insulator between
the desoldering tip and the solder joint. If the dirt is not removed,
additional heat will be needed to melt the solder. This added heat
can cause damage to the board. (Refer to incident #2 on page 78.)

Consideration should also be given to adding solder to a connection if there is very little solder on the joint or if it is old. The solder
connection is what transmits the heat through the hole to the component side. If there is very little solder there, it will be next to impossible to get the heat through to the component side to melt the solder.
So add some more solder. As far as the old connection is concerned,

the solder joint is not the same as when it was first made. The joint becomes primarily lead rich due to the leaching of the tin into the copper. Adding fresh solder will make it much easier to remove.

The main items to remember are: the proper use of heat for the mass of the joint, the right-sized tip, and lack of pressure when desoldering. This all minimizes the amount of time needed to desolder.

As you can see, the same criteria apply to desoldering as apply to soldering:

- Cleaning
- Flux
- Time
- Mass
- Temperature

DESOLDERING STATIONS

There are a variety of powered vacuum desoldering stations on the market. Each has its own peculiarities and design. Some are far easier to use, maintain, and change tips than others. To start with, the handle should be comfortable to hold, with the fingers of your hand in total control of the tool and as close as possible to the desoldering tip. This ensures accuracy and provides greater control to the operator. The closer your fingers get to the tip of the tool, the better the work will be. Try writing your name holding the pencil by its eraser end—the same thing applies here.

Station Maintenance

The collection chamber would be better if enclosed to protect the operator from possible burns. The operators should not have to be concerned about burning themselves, which could happen where the chamber is externally exposed. The chamber has hot solder that has accumulated at a high temperature. A chamber outside the handle does not allow the operator's fingers to get very close to the tip.

Cleaning of the collection chamber should be easy. Keeping it clean ensures that it will operate properly. One suggestion is to put a thin film of pure mineral oil on the inside of the glass tube and on the baffle if it has one. Make sure the same filters are used as were originally supplied with the unit when it was purchased. Cotton batting does not effectively filter out contaminants in the same way as a heavy felt filter. Contaminants getting past this filter end up damaging the motor operating the pump.

Changing Tips

Changing tips, which is a major consideration (one size cannot do all joints), should be done easily—that is, loosen a set screw, pull out the tip, install the proper one, and tighten the set screw. All you need is a simple screwdriver, no wrenches or other tools that are needed for some other systems.

From experience over the years, pistol grip-type desoldering tools do not appear to give the same fine control as other types. A person's wrist does not have the flexibility and control that the fingers have for performing a desoldering task.

SURFACE MOUNT

For desoldering surface mount components, the process is a little different from THM. First of all, greater care is a must. Through hole mount components can sometimes be a little forgiving because the plated through hole and the pad on the component side help dissipate excessive heat. Not so with surface mount—there is *no* forgiveness. Too much heat or time and not only will the lead and solder be removed from the board, but the footprint as well. The connection areas are only glued to the board, and as noted earlier, that glue loses 80 percent of its strength when heated. The idea with surface mount is the right tool, the right temperature, the right time, and the right technique.

Handpiece Tips for Desoldering SMT

There are a very wide variety of tips and handpieces that can be used for removing SMT items. For example there are bifurcated, tunnel and four-sided tips that will remove capacitors, SOTs, a variety of SOICs and surface mounted sockets. A tweezer type handpiece with the proper tip will also remove a number of these items as well. Hot air can also be used to remove a number of these parts. For PLCC/SOJ, the preference is the tweezer tool with the appropriate tips or the large piece of equipment mentioned under method 4 on installation of PLCC/SOJ. For the large QFP shown in Figure 6–2, a different type of handpiece with a tip that will match the outline of the QFP and a vacuum cup in the middle is preferred, unless you have the large machine as indicated in method 4 on dealing with the installation of these items.

Technique for Removing SMT Items

One special problem that arises with surface mount components/ devices concerning removal relates to items that are glued to the surface of the board. Gluing keeps items in place during the manufacturing process. Most of the time you will not have any idea whether there is adhesive involved until you try to lift the item from the board after the solder has melted. So you go back and try again because "this thing has got to come off." This is where damage occurs. The board and heated areas have not had a chance to cool off, and the glue holding the footprint(s) has not regained its strength. Wait until the area has cooled. When **all** of the solder has completely melted on your next try, give the component/device a very, very gentle tweak. This will break the adhesive bond holding it in place. This applies whether the component is large or small.

MANUFACTURING EQUIPMENT

When it comes to machines used for manufacturing surface mount boards, there are usually four types used:

1. Vapor phase uses liquid vapors to reflow the solder paste that is applied to the footprints.

2. Infrared, which is somewhat similar to an ordinary oven, is also used to reflow the solder paste that has been applied to the board.

3. Convection heating is another method for reflowing solder paste and is receiving wide acceptance in the industry.

4. Wave soldering machines have been used for years for THM. When using this machine for SMT, adhesive is used to hold each of the surface mount items in place on the bottom side of the board (solder side). This is where the technique described for desoldering SMT items is an absolute necessity. The operator/technician probably will not know how the board was made, so this is a good habit to get into for surface mount.

For surface mount desoldering tools, a separate catalogue would be needed to show half of them. They are similar in most ways to the soldering tools for SMT. Hot-air hand tools and systems are very popular. Tweezer-type tools for removing nearly every type of component or device are also available. Each person decides how comfortable he/she is with a particular piece of equipment.

NOTES

CHAPTER 8 Evaluation

Objectives

After studying this chapter, you should be able to:

- List the characteristics of a circuit board, including substrate material; single, double, or multilayered; pad sizes and shapes; widths of tracks; locations of ground and voltage planes; methods of coating removal, if required; type(s) of lead terminations.

INTRODUCTION TO BOARD REPAIRS

Equipment being worked on will have to be checked out to find the problem. Once that has been established, unplug the unit to avoid losing a finger (a man's ring finger was blown off when his wedding ring came into contact with some high voltage). Remove chains, rings, watches, and any other metal jewelry. Following proper safety precautions will leave you alive and in possession of all your fingers.

Some repair depots have the feeling that if they make a mediocre repair, the unit will be returned shortly, and another repair made, and another bill given to the customer. Customers are smarter than this. They want the repair done right the first time so that they do not have to take the unit back again.

A repair service increases its business by providing quality repairs, not the type that will break down easily. A service manager once said, "We don't care how it's done. We just want to get it out." Would you take your equipment there?

It doesn't take a lot of time to do the job correctly. For the job to be effective, you need to know how it should be done.

There is a sequence of events that should take place whenever repairs to a circuit board are to be made. Some of them are done, others are not considered because no one has ever told them what needs to be done. One important area is the evaluation of the circuit board by technicians prior to deciding what tool they should pick up before doing any work on the board.

EVALUATION OF THE BOARD

The evaluation should consider all aspects of the board to ensure that nothing is overlooked. Someone who has received a board to be repaired should start by basically performing a very simple task. Look at the board and determine just exactly what problems will likely be encountered, the procedures to be used, and the type of repair that will be required.

Circuit Board Evaluation

To find out what has to be done, closely examine the board to find out what was involved in its manufacture. Consider:

- The material the board is made from—i.e., fiberglass/epoxy, plastic, phenolic resin, ceramic, Kevlar, nylon, or any other materials.

- The thickness of the board: standard, very thick, very thin.

- Whether the board is single-sided, double-sided, or multilayered.

- If it has plated through holes.

- The pad sizes and shapes.

- Width of the traces/runs/tracks.

- Ground or voltage planes that could cause problems.

- If a coating is on the board. If there is, you now have an additional concern.

- Lead terminations, full clinch, semiclinch, straight through; spaded or offset pads on older boards.

If any of the above items are disregarded, you can easily end up with unnecessary repairs.

This may sound like a task for people who have never done work on a circuit board before, but a board can easily be destroyed. Components can be replaced at a fairly reasonable cost, but not the board. Without the board, all you have is a handful of spare parts and possibly a not so happy supervisor, boss, and/or customer.

Before touching a tool, think about it first. It doesn't take long. Make sure you know what you have in your hands and what you are going to do with it to make a good, long-lasting, reliable repair.

NOTES

CHAPTER 9 Conformal Coatings and Solder Masks

Objectives

After studying this chapter, you should be able to:

- Describe the best method for the removal of a single part polyurethane-type coating, without damage to the circuit board.

REMOVING COATINGS

To find out why a unit is not working, removal of a coating or mask may be one of the first things that has to be done, especially if the test points are covered.

Some coatings are relatively simple to remove, others can be very difficult, especially if you try to use a soldering iron to remove them. A soldering iron is for soldering and that is all. Another tool is needed to remove the coatings.

Be careful what you use to remove coatings. (It is unbelievable what some people will use.) It has been recommended that a chemical be used to remove epoxy-type coatings. When asked what material their board is made from and they respond with epoxy fiberglass, technicians that have used chemicals start to think: If the chemical removes the epoxy coating, what about the epoxy material in the board?

EPOXY COATINGS

One of the easiest methods to remove epoxy *without board damage,* is to use a hot-air jet, where the heat can be precisely controlled. A large flow of hot air is not required if you are working on a small area. It could also be hazardous in regard to hand and finger burns. (A heat gun is not the tool. You need the right equipment.) Apply a stream of hot air to the area and use a wooden stick to chip and lift off the overcured epoxy. Make sure it is completely removed from both sides of the board and a small area around the solder connection. This material is an insulator and there will be difficulty removing the solder if the coating is still there. The wooden stick prevents damage to the board, which would happen if an instrument such as a screwdriver or sharp knife-type implement were used. If you damage the board, you now have another area to repair.

Some solder masks are an epoxy-type material. The method just discussed can also be used for removal of this type of mask.

POLYURETHANE COATINGS

Another coating that can be removed using hot air is in the polyurethane group. This coating will become liquid very quickly when heated and has to be pushed away from the solder connections or it will flow back over the area. For a single-part polyurethane coating, this method is fine. For the two-part polyurethane, the procedure is the same as for the epoxy-type coating. Again make sure that all of the coating is removed before trying to desolder. A hot-air system is the best procedure.

OTHER COATINGS AND MASKS

There are other coatings and solder masks that are relatively simple to remove, and there is one that is a real problem.

Acrylic Coatings

Easier coatings, such as acrylic lacquer, can be removed with something as easy as a mild solvent, such as isopropanol (IPA). Just concern yourself with the area where the desoldering or repair is to be done. Why remove the coating from an area you are not concerned with and which is not part of the required repair? It now becomes an area you have to remember to recoat after the repair is completed. The coating was put on the board for a purpose. Put the coating back on to provide the same protection the components and board had before you removed it. If the new coating is not the same as the material you removed, it will have to be checked to see if it is compatible with the coating already on the board. Some coatings do not interact well and should not be brought into contact with each other.

Varnish Coating

In some cases you may be working on older types of boards and coatings. One of these coatings may be varnish. If the board has a light brownish covering and it appears that some of it has flaked off, the possibility is good that it is varnish. This could be removed with a strong solvent, but once again consider the material of the board first. The safer method is the hot-air system and a wooden stick. If you happen to smell something resembling linseed oil, you are just confirming that the coating was indeed varnish.

Parylene Coating

Coatings can also be removed using motorized abrasive tools. A light solder mask can be very easily removed with one of these tools along with a very gentle touch. By the same token a tough coating such as Parylene has to be removed in this manner. You have no choice. Extreme care must be taken when coping with this coating to prevent board damage. It can happen very quickly because the coating is so very thin—i.e., in the area of 0.001" to 0.002" (0.025 mm to 0.05 mm).

When it comes to coatings or solder masks, it takes great care and caution to remove them effectively. Practicing different methods on various coatings would be a big help.

Just remember that every last bit of coating or solder mask must be completely removed from all of the area that will be affected by your work. The material can act as an insulator to your desoldering tool. If it burns, it could become toxic; in any case, it will smell very unpleasant.

PREPARATION FOR PART RE-INSTALLATION

It is recommended that after removal of the coating and the part, each hole where the component was removed should be filled with solder to remove any conformal coating particles. They may have ended up in the holes. Desolder the holes, install the part and solder the leads. Thoroughly clean the area and re-apply the coating.

CHAPTER 10 Pad and Track Repairs

Objectives

After studying this chapter, you should be able to:

- Describe the preferred method for reinstalling a lifted pad back onto a single-sided board.

- Describe all of the tasks involved in replacing a badly damaged pad and track on a double-sided board where the overlap repair is only two pad widths away from the place where the badly damaged pad was originally installed.

INTRODUCTION TO PAD AND TRACK REPAIR

Pad and track repair is common, probably too common. It becomes necessary because of inexperience or lack of proper desoldering tools and poor technique by the person(s) doing the job.

The repair can be done fairly easily. As long as the individual knows what has to be done, the repair does not take a great deal of time either.

For reliability, the pad should be fastened back down onto the board with a suitably sized eyelette or funnellette. For a single-sided board, an eyelette is the better choice because it provides a good mechanical connection to the component side of the board when it is flared out. It does no good at all to just glue the pad back down to the board. The heat that will have to be applied to solder the connection will simply lift the pad again. Gluing is obviously not an effective repair method.

SINGLE-SIDED PAD AND TRACK

A single-sided board with a lifted pad is repaired by placing the correctly sized eyelette through the pad and into the established hole or if necessary an enlarged hole. The portion that extends through to the component side is then swaged down flat onto the board surface. The component lead is then inserted and soldered into place. Now you have a reliable repair.

DOUBLE-SIDED PAD AND TRACK

A double-sided board repair of a lifted pad should be done using a funnellette rather than an eyelette. The funnellette provides the most reliable type of repair for a double-sided board. But if only eyelettes are available, they will also make an effective repair. (See Figure 10–1.)

With the funnellette the tube portion is put through the hole after first going through the lifted pad. The flared part is held in place while the portion sticking through the board is swaged to hold it in place, as shown in Figure 10–2. The funnellette is then soldered to the

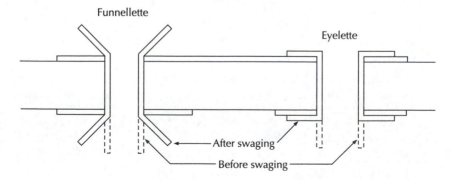

FIGURE 10–1 Funnellette/eyelette installations

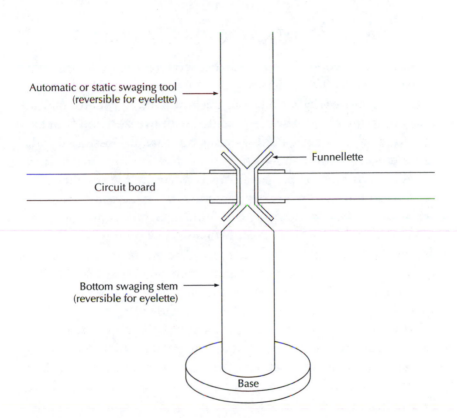

FIGURE 10–2   Swaging of a funnellette

pads on both sides of the board prior to the lead being installed and soldered. This is done to ensure a good connection between the pad and the funnellette.

With the eyelette the tube portion is put through the pad and held. The flat part is pressed up against the lifted pad and board. The piece sticking through the board is then swaged flat to secure it in place. The lead is installed through the eyelette, terminated, and soldered into place. The soldering process should include the flow of solder from the lead to the eyelette to the pad, on both sides of a double-sided board. This provides a good, reliable connection.

PTH PAD AND TRACK REPLACEMENT

Should the lifted pad and track be severely damaged and need replacing, the method differs slightly. The damaged pad and track are carefully removed by scoring the track at a point where it is still attached to the board. The piece is lifted with tweezers and bent back and forth at the scoring until it breaks off. If at all possible this distance down the track should be at least two and one-half to three times the width of the pad area. This is to ensure the stability of the track repair joint so that it does not reflow when the component lead is soldered back into the eyelette or funnellette. If this distance is not possible, cover the overlap repair with an epoxy coating. This will act as a heat sink to protect the connection. The overlap of the new track onto the old track should be a minimum of $1/8$" (3.18 mm) to a maximum of $1/4$" (6.35 mm) for a good repair; it does not matter what the width of the track might be, as shown in Figure 10–3. Be sure that the length of the track between the pad and where the track is being soldered to the old track is down flat on the board. The remaining process is the same as though there were just a lifted pad.

Soldering the Overlap

When soldering the overlap into place, a wooden stick is placed on the new track where it meets the end of the old. The old track should be tinned before installing the new pad and track. Flux the area.

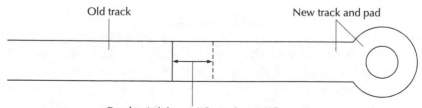

FIGURE 10–3   **Pad and track repair**

Apply the soldering iron right next to the wooden stick. When all of the solder melts, move the iron down the new track towards the end of the overlap. Move the stick up onto the top of the overlap. Take the iron to the end of the new run and off at right angles to the track. This prevents the soldering iron from drawing the solder away from the overlap area. Remember, the solder goes towards the heat so it will follow the iron down the track. Keep the wooden stick on top until the solder solidifies—lightly, very lightly. (See Figure 10–4.) If pressure is applied, the end of the new track will lift up from where it is supposed to be lying. It's time to clean up—the job is finished.

SMT PAD AND TRACK REPLACEMENT

For this type of repair, the only thing you can do is to rebond the pad and track to the board. A two-part anerobic adhesive can be used if

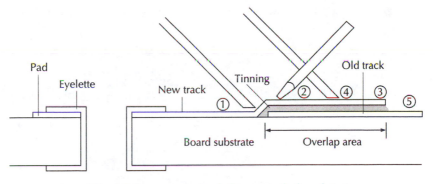

- Wood stick presses new track down onto the board starting at the eyelette and moving to point ①.
- Iron is applied next to wood stick to melt tinning previously applied at point ②.
- Iron is moved to point ③.
- Wood stick moves lightly to point ④.
- Iron moves to point ⑤ and is removed.

FIGURE 10–4   New pad and track repair

the pad and track are still serviceable. If not, a new pad and track will be required. Kits for this type of repair are available which allows you to use your iron with a special tip, to bond the pad and track to the board. Don't forget to read the instructions with the kit.

CHAPTER 11 Run/Track/Trace Repairs

Objectives

After studying this chapter, you should be able to:

- Describe the complete repair of a track where the damage involves an area 1/2" long and a piece of tinned copper track is to be used.

- Describe the complete repair of an irregularly shaped track by using the track from another board of the same type to effect the repair.

RUN/TRACK/TRACE REPAIR

Run/track/trace repair is relatively easy. There are two dependable methods that can be used, considering the materials that you may have in order to complete the repair.

PREPARATION OF THE DAMAGED AREA

The first thing that should be done is to remove the damaged piece of track. This is done by scoring a straight line across the damaged run at a point where it is still attached to the board. If the scoring can be done at the point where the damaged track is removed, rather than a straight cut, a bevel cut of approximately 45° will make it easier to install the new track. The damaged track is lifted to the mark made across the track and moved back and forth until it breaks off. Do this for both ends. If the board in the area of the opening has exposed fiber or is otherwise damaged, this should be repaired first. If it has been burned, clean off the burned material and replace it with an appropriate compound. The burned material can act as a carbon resistance. If it is an epoxy fiberglass board and the fibers are exposed, after the area has been cleaned out, reseal the spot using an epoxy compound to make it compatible with the remainder of the board. Allow the epoxy to cure properly and smooth it off using very fine (600 grit) emery paper. Make sure this is done before the track repair is made. Moisture coming into contact with the fibers will cause further problems later on because the moisture tends to travel along the weave of the fiber, which could result in blistering of the board.

FLAT TRACK REPAIR

The flat track repair can be done using either pieces of track from a similar type of board that has been scrapped or from commercial pieces of plain or tinned copper. The use of track from a similar board in most cases is the best idea, especially if irregularly shaped runs are to be repaired. (So it is not a good idea to throw away all of your old boards.) To remove a piece of track from an old board, just cut across the run and apply heat at that point. This overheats the adhesive under the track and permits the track to be lifted easily. Do not forget to remove the old adhesive from the back of the piece of track at each end. Otherwise you will not be able to solder the overlap portion of the track.

If the damaged run has a break of less than 1/8" (3.18 mm), the piece of repair material can be placed over the break and should extend a minimum of 1/8" (3.18 mm) on each side of the break to a maximum if 1/4" (6.35 mm). The track being repaired should have a very small amount of solder applied onto either side of the break on the old run. This is done by first cleaning off either end, adding liquid flux, and tinning the two areas. Clean off the used flux and add more liquid flux. Place the new piece of track over the area to be repaired. Tack one end first, then fully reflow solder the opposite end. Go back to the end you tacked and reflow solder it onto the old track. A wooden stick should be used to hold the track in place at either end while the solder is solidifying. Clean off all of the flux residue.

When the break is in excess of $1/8"$ (3.18 mm), the new piece of track should be formed to the board, as shown in Figure 11–1. In this case, it is recommended that the track being repaired be cut at a 45° angle. It will be easier to form the new piece of material into the area where the track is removed. After pretinning both ends of the old track, the new piece should be fluxed and tacked at one end. Then using a wooden stick, form the new track down onto the board and along to the opposite end of the break. The end of the new trace is fluxed and reflow soldered into place. The tacked end is then reflow soldered. The stick is used in the same manner as the overlap for the pad and track repair in Chapter 10. See Figure 10–4.

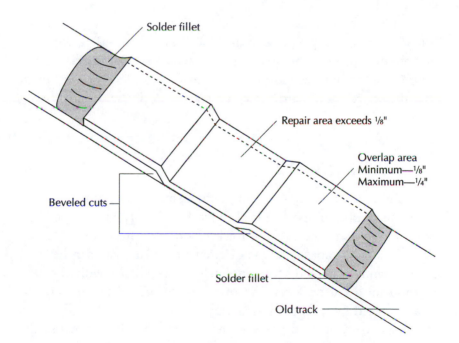

FIGURE 11–1   Flat track repair

For a track repair that is more than ¹/₄" (6.35 mm), not only should the track be formed to the board but also fastened using an adhesive to bond the track to the board. A conformal type coating could be used to cover the complete area of repair.

Commercial replacement track is available in a few forms:

- Tinned pieces of copper of specific length
- Plain copper tape with an adhesive backing

The tinned copper is very easy to use. All you have to do is cut it to the length needed. It is usually only an inch or two long, so it is not suitable for extended repairs.

Where a lengthy repair is required, the copper tape is preferred. The adhesive will keep the trace in place on the board. The entire length of the copper tape will need to be tinned. The new tape butts up against the old trace, and a normal flat track repair is done at either end.

On RF circuit boards, make sure that there are no sharp bends or folds when effecting repairs; otherwise they will interfere with the normal operation of the board. That is why a similar unserviceable board would come in handy. Just remove the piece needed from the unserviceable board and reinstall it on the board being repaired.

BUS WIRE REPAIR

A second method of repair is simply by using a piece of bus wire or even a piece of a component lead to fix the track. The wire should be a minimum width of half the track in order to have the necessary current-carrying capacity as the original track. The wire can be used as is, slightly flattened, or completely flattened. As with the first method for a flat track repair, both ends of the repair area are cut away, cleaned up, and pretinned. As the wire can not be formed to the board because of its rigidity, it can be laid directly across the break and the tinning reflowed onto the wire. Be sure that the copper end of the wire is covered.

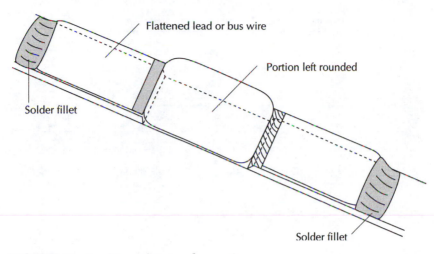

FIGURE 11–2   Bus wire track repair

It is more difficult to hold a round wire in place onto a rounded surface. By slightly flattening the wire with flat-nosed pliers (Figure 11–2)—not serrated needle-nosed pliers—it will be easier to hold the wire in place while the reflow soldering process is being performed. Partial flattening of the wire at either end of the repair piece, allows the flat portion to lay down on the old track and the still rounded portion of the wire, to fill in the space between the two ends of the break.

FINE LINE REPAIR

For very thin traces that have to be repaired and are not completely severed, another method is available. Small wires can be used to make these repairs. The old trace is tinned, and the used flux is removed. New flux is added, and the repair wire is put into place with a pair of tweezers. The clean soldering iron tip is applied at one end of the wire as shown in Figure 11–3. The solder melts, and the solder tension will align the wire straight down the trace. Clean off the board. The repair is complete.

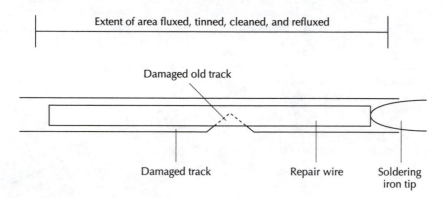

FIGURE 11–3   **Fine-line track repair**

CHAPTER 12 **Burn Repairs**

Objectives

After studying this chapter, you should be able to:

- Describe all of the details concerning the repair of a board that has burn damage caused by a vertically mounted high-wattage resistor that has been improperly installed. The board is burned completely through.

BURN REPAIRS

Burned circuit boards are usually the result of an improperly installed component—most of the time due to a resistor. (See Part I, Chapter 6.) There can be other causes, but resistors are the most common. Regardless of how it occurred, the damage must be repaired.

PARTIAL-BURN REPAIR

If the burn does not go all the way through the board (Figure 12–1), clean out all of the carbon material and slightly bevel the area inwards in three or four places under the opening. Prepare a mixture of epoxy and fiberglass powder; the powder should be approximately 10 percent of the total. Place the mixture into the cleaned out area.

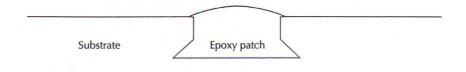

Substrate   Epoxy patch

FIGURE 12–1   Partial-burn repair patch

The powder is added to give a flexibility to the mixture that would not otherwise be there. Epoxy on its own is very rigid and brittle. Make sure it comes above the top of the board because the mixture will contract slightly when it cures. The excess epoxy is removed after curing. The area made smooth by a combination of a motorized abrasive, emery paper, and a very gentle touch. Any track or other item can be reestablished.

THROUGH-BOARD BURN REPAIR

Where the burn goes all the way through the board (Figure 12–2), drill and clean out all of the burned material. Using a motorized abrasive, bevel both sides of the burned out area so that the epoxy and fiberglass mix will have a larger surface area to adhere to and an anchor in the middle of the board. Place a block of Teflon or similar material under the cleaned out hole so that the epoxy will not stick to it. Put this on one side of the board and hold it in place with static dissipative tape or a small clamp. Fill the hole completely with the epoxy mix, again making sure that it comes above the top of the board. After curing, the excess epoxy is removed. Everything else that is required to bring the board back to its original status should now be done. This could include drilling out a hole within the patch that was made, installing an eyelette or funnellette for a double-sided board, and replacing pads, tracks, etc.

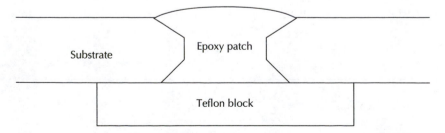

FIGURE 12–2   Through-board burn repair

PLUG-TYPE BURN REPAIR

For burned out areas of ¹/₂" (12.7 mm) or larger, a piece of board stock or a similar board should be cut to fill in the cleaned out area, as shown in Figure 12–3. The sides of the new piece of board are then beveled, ordinary epoxy is placed around the edge of the hole, the piece is installed in the hole, and a Teflon block is taped or clamped to the board. After curing, remove the block, level off both sides of the board (if necessary), and reinstall, reconnect, drill, or whatever else is necessary to bring the board back to serviceable condition.

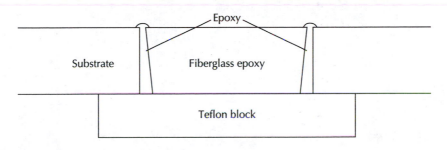

FIGURE 12–3   Large burn repair

NOTES

CHAPTER 13 Edge Connectors/ Fingers

Objectives

After studying this chapter, you should be able to:

- Describe the complete procedure for the removal and rein-stallation of a badly damaged edge connector/finger.

REPAIRING EDGE CONNECTORS/FINGERS

The repair of edge connectors/fingers may turn out to be rare, but what do you do if it needs to be done?

Different repairs may be required, from simple cleaning to replacement of badly damaged edge connectors.

CLEANING OF EDGE CONNECTORS

When cleaning a connector/finger is required, one of the most effective means is the use of a solution of 80 percent alcohol and 20 percent water. Soak, dry, and polish the connector(s). Ammonium hydroxide is considered the best cleaner, but it requires the extensive use of rubber gloves, apron, boots, goggles, and respirator. The second-best cleaner, the IPA and water solution, is much easier to use.

A popular idea for the cleaning of the edge connectors is the use of a pink pencil eraser. This is not a good method. It not only leaves a gum residue on the surface of the finger, which must then be removed, but it also removes the already very thin layer of gold. The gold is there to provide good conductivity. If the gold is gone, so is the conductivity—and the connection.

LIFTED FINGER REPAIR

A second type of repair may involve reinstalling a partially lifted finger. An adhesive suitable for this task should first be obtained; some type of anaerobic adhesive would be best. A glue containing cyanoacrylates is not recommended. If this type of adhesive comes in contact with a hot soldering iron, the gas produced is considered to be very toxic.

Apply the adhesive as directed on the container between the lifted finger and the circuit board. Clamp or tape (static dissipative tape) the finger to hold it in place while it cures. After the adhesive has cured, remove the clamp/tape and any residual adhesive that may have ended up on the finger. Do this very carefully so as not to remove any of the gold. Stay away from using a motorized abrasive when doing the cleaning. The frictional heating could cause the adhesive to overcure, resulting in the finger lifting up again.

EDGE CONNECTOR REPLACEMENT

For the badly damaged edge connector that has to be replaced, the following procedure is suggested. (See Figure 13–1.)

1. Remove the edge connector by scoring the point where it is still fastened to the board. Lift it up and move it back and forth until it breaks off.

2. Clean the board to remove any glue that may have been left behind by the old finger.

3. Obtain a new finger by removing one from a similar board, from something commercially available, or by making one from scrap copper.

4. Solder the finger onto the old track in the same way as a flat track repair. (See Chapter 11.) Overlap the old area 1/8" (3.18 mm) to 1/4" (6.35 mm) and reflow the solder that was added to the old track. Make sure that the finger is properly aligned.

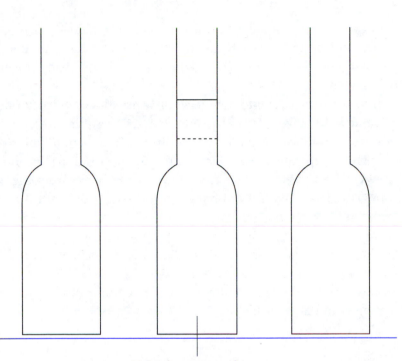

New edge connector finger
showing overlap onto old track

FIGURE 13–1   Edge connectors/fingers

5. Glue to the board using the same type of adhesive used to repair the partially lifted finger. Allow to cure, then clean.

Where it was necessary to use copper or a tinned commercial finger when replacing the connector, the connector will have to be plated with gold in order for it to perform as it should and be compatible with the other edge connectors/fingers.

PROBLEM EDGE CONNECTORS

It has been found that when edge connectors become partially damaged by a scrape or improper connector installation, it will depend on how far the damage goes into the connector/finger as to what needs to be done for repairs. The fingers normally comprise the copper base, covered by nickel (referred to as a lock-off) then topped off by the gold.

If the damage goes into the nickel or worse, down to the copper, a galvanic action takes place between the metals resulting in rock chloride crystals ending up on top of the finger. These "rocks" are strong enough to separate the contacts and cause operational problems. They can be knocked off by just removing the board from the edge connector and when it is put back in, the board works just fine. The problem however, has not been fixed as the crystals will grow back again.

The connector will need to be replated to completely fix the problem. If the damage is down to the copper, the nickel plating will be required before the gold. If the damage just goes to the nickel, gold plating is all that is needed.

CHAPTER 14 Modifications

Objectives

After studying this chapter, you should be able to:

- Describe the procedure required for the proper installation of a modification wire from a resistor to a DIP and from a DIP to a flat pack/gull wing-type lead.

MODIFYING CIRCUIT BOARDS

When engineering changes necessitate one or more modifications to be made to a circuit board, the new wiring should be done on the component side of the board. The wire for the change should be installed by running it on the x- and y-axis of the board without going underneath components. An adhesive compatible with the board substrate should be used, again staying away from glues with cyanoacrylates. Something as simple hot melt glue does a good job. This should be placed at each bend and every inch (25.4 mm) for the entire length of the modification wire.

The connections at either end of the wire will depend on what the wire is going to be connected to at its termination point.

RESISTOR OR SIMILAR TERMINATION

If the termination point is a resistor or similar type of round lead, the wire—after being tinned—is wrapped around the lead from 180° to 270° in the same manner as a wire on a turret terminal. (See Figure 14–1.) Put a small amount of solder liquid flux on the wire on the component side of the board as well as on the bottom of the board before soldering.

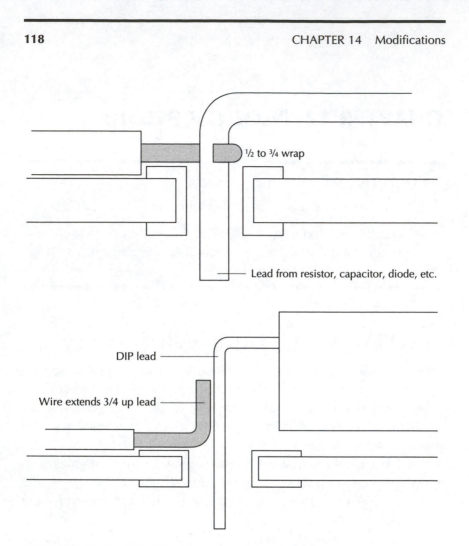

FIGURE 14–1   Modifcations for through hole mount

DIP TERMINATION

If the lead belongs on a DIP, the wire goes to the edge of the pad and is then bent up the lead, not down through the hole. It should extend approximately halfway to three-fourths of the way up the lead. The insulation on the wire ends at the edge of the pad where the wire is to be terminated. Add liquid flux in the same way as the resistor installation.

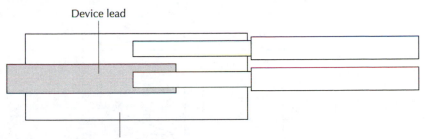

FIGURE 14–2   Surface mount "mod" wire installation

FLAT PACK/GULL WING TERMINATION

For the flat pack or gull wing-type of surface mount lead, the wire is placed either on top of or beside the lead and soldered into place, as shown in Figure 14–2. Again the wire insulation terminates at the edge of the pad area.

By installing and soldering the wires on the component side, the height of the components protects the wires from being accidentally ripped off by being caught on another board or equipment case. This is a much safer procedure.

CHIP CAP/RESISTOR TERMINATION

For a passive component attachment by a mod wire, it should enter onto the pad and termination area at a right angle and lay across the width of the termination. See Figure 14–3.

One final suggestion: After the repair has been completed, closely examine the circuit board to see if there are any scratches, gouges, or other damage that may have exposed some copper. Where there is, clean it, flux it, and if necessary add solder or just reflow the solder or tinning already present, and cover the copper. This will prevent further oxidation and deterioration of the area, and maybe a repair being needed at a later date.

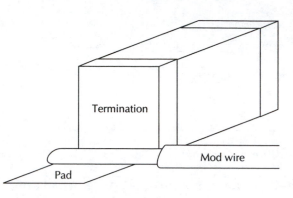

FIGURE 14-3   Passve Component

Based on the previous information and realizing that many repairs will not be able to be done exactly as has been indicated, the methods closest to it will give you a good example of what is required. A little thought will help you make the final decision on how the repair is to be completed. By performing reliable repairs, you will enhance your reputation for good work.

APPENDIX A Electrostatic Discharge (ESD)

If electrostatic discharge (ESD) is not properly dealt with, all of the effort put into other areas of concern goes for nought. It is important to know the fundamentals.

1. The boss should know that ESD causes damage and to know how it occurs. The boss should also be informed about how to control ESD and keep it at manageable levels. If he/she does not know—or even worse, does not care—you will never be able to adequately deal with your problem.

2. Everyone in the plant, from shipping and office clerks, should be made fully aware of why precautions must be taken in the operation and proper use of equipment and materials.

3. *For personnel working on circuit boards only,* equipment and material need to be obtained, and the workbench and everything on it need to be grounded. This includes all the electrical units, mats, soldering irons, flooring, and wrist straps, to mention a few. Ionizers should be used if insulative materials are required near the work area. Workbenches and mats should be static dissipative and used in conjunction with a grounded chair covered with antistatic fabric.

4. *For personnel working on live high-potential equipment,* the setup changes: A solid maple bench with no finish, an insulative floor mat, antistatic fabric on the chair but no internal ground path, and safety notices regarding the fact that jewelry, wriststraps, and other conductive items should not be worn.

5. Packaging, in-plant transportation, handling, and storage of parts and boards have to be addressed. Steps need to be taken in these areas to prevent ESD damage.

Why buy a wriststrap and bench mat if nothing else is done to prevent ESD? The money spent for ESD can be money down the drain if nothing else is considered. Yet that is what some people do, and they feel that is all the protection they need.

With the sensitivity of the present IC packages and the ultrasensitive ICs now being produced, get used to the idea that ESD has to be dealt with if your business is to survive. (Study Tables A–1 and A–2, Figures A–1 and A–2.)

A few years ago a major computer manufacturer was concerned because its latest IC would be totally destroyed with as little as 25 volts of ESD. People do not feel static until it gets to the 2,500 to 3,500 volt range, so they would have absolutely no idea they had just destroyed one of these ICs.

TABLE A–1   Chart showing the polarity of the charge generated when items separate from one another

Sample materials	Charge generated, + or –
Asbestos	
Acetate	
Glass	
Human hair	
Nylon	
Wool	
Fur	
Lead	
Silk	
Aluminum	
Paper	
Polyurethane	
Cotton	
Wood	
Steel	
Sealing wax	
Hard rubber	
Acetate fiber	
Mylar	
Epoxy glass	
Nickel, copper, silver	
401 epoxy resist	
UV resist	
Brass, stainless steel	
Synthetic rubber	
Acrylic	
Polystyrene foam	
Polyurethane foam	
Saran	
Polyester	
Polypropylene	
PVC (plastic)	
Kel F	
Teflon	
Viton	
Silicone rubber	

Increasingly Positive

Increasingly Negative

TABLE A–2   Voltages generated from specific incidents dependent on the humidity level

Event	Relative Humidity	
	10%	*55%*
Walking across carpet	35,000	7,500
Walking across vinyl floor	12,000	3,000
Motions of benchworker	6,000	400
Foam chair	18,000	3,000
Tape from board	12,000	3,000
Triggering solder pullit	8,000	2,000
Cleaning with eraser	12,000	2,000
Freon spray	15,000	6,500
Removing DIPs from plastic tubes	2,000	400
Removing DIPs from vinyl tray	11,500	2,000
Removing DIPs from styrofoam	14,500	3,500
Removing bubble pack from PWBs	26,000	7,000
Packing PWBs in foam-lined box	21,000	5,500

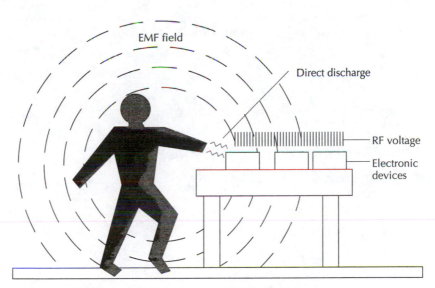

FIGURE A–1   Various occurrences having an effect on electronic devices

*You can't hide from static—
it will get you in the END.*

FIGURE A–2

NOTES

APPENDIX B   The Workstation

If workers are happy with their surroundings and reasonably comfortable, they will do a far superior job than if they are in a dull, gloomy, antiquated and dirty workhouse atmosphere.

To start with, let's look at a few of the basic needs.

1. *Lighting*—This item for some reason tends to be ignored. How can people produce a high-quality product if they cannot see what they are supposed to be doing? Soldering operations should be performed on a bench with the light level at 100 footcandles or more diffused over the entire work surface. Anything less is inadequate, and employees end up with eye strain, eye checkups (which require time off), and eyeglass purchases. When this type of situation is present, people cannot work effectively. It takes very little to remedy this condition. A combination of fluorescent and incandescent light provides the best lighting.

2. *Seating*—Ergonomics is something companies need to become acquainted with, especially as it applies to electronics industry workers. Make it a good chair—not an easy chair, but something the buttocks can comfortably adapt to.

3. *Workbench*—It should be easy to clean and maintain, keeping in mind always that ESD is a constant concern. The more time spent maintaining the bench, the less time devoted to production. Shelving should be provided for test equipment or other items that are not needed on the bench. Provide racks, bins for parts and supplies, variable surface heights if required, footrests, drawers, and anything else needed to keep the surface available for the items being produced or serviced. (See Figures B–1 and B–2, and Table B–1.)

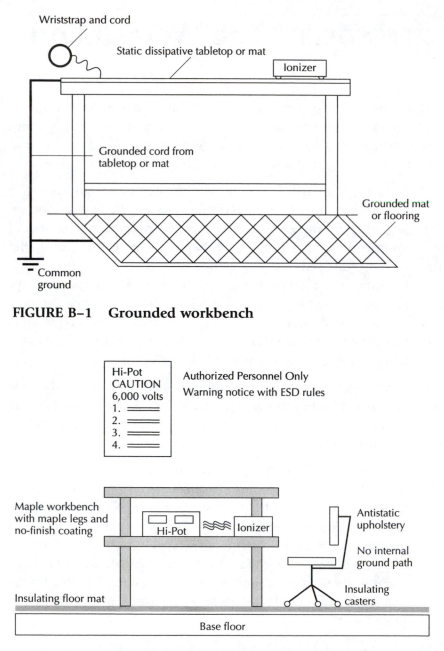

FIGURE B–1   Grounded workbench

FIGURE B–2   High-voltage workstation over 250 V

TABLE B–1   High-voltage workstation over 250 V

- All wood construction including legs and risers.

- No surface finish except oil rub.

- No static dissipating laminate or mat.

- Minimize use of insulating plastic.

- Use topical antistatic treatment or ionization to control static charge on plastic surfaces.

- Use insulating mat that meets the requirements of ANSI/ASTM D178 Type 1 Class 2 on the floor.

- Use static dissipative shoes (not conductive).

- Do not use heel or toe grounders.

- Use antistatic chair fabric with insulating casters.

4. *Air*—The quality of air sometimes has to be strained to provide a reasonably healthy environment. Check air quality periodically so that personnel will not suffer any ill effects while doing their work. With the types and variety of chemicals being used these days, this is a must.

5. *Equipment*—The cheapest is rarely the best or the most reliable over the long term. A worker uses the equipment for eight hours a day, so it must be easy to maintain and operate, change parts quickly, modify its opera-tion quickly, and take up as little room as possible on the bench surface. The time it takes to clean equipment or change operations (tips and temperature) translates into lost time. Even though a piece of equipment may be expensive to purchase initially, it will actually cost less over the years because of its ease of use. Check equipment thoroughly before you buy. Get a stopwatch and check how long it takes to clean or change

hot tips. Then start multiplying. Equipment is important but far less so than the first four items mentioned here. Workers cannot operate the equipment if they are not at work.

The less time taken off by workers due to sickness or related appointments, the more time they will have to produce, which is what they are being paid to do. Whether it is eye fatigue from poor lighting, back problems from poor seating or workbenches, or respiratory concerns due to air quality, the result is lost time and reduced production. These cost your company.

Index